I0074391

Logic with a Probability Semantics

Logic with a Probability Semantics

Including Solutions to
Some Philosophical Problems

Theodore Hailperin

Published by Lehigh University Press
Co-published with The Rowman & Littlefield Publishing Group, Inc.
4501 Forbes Boulevard, Suite 200, Lanham, Maryland 20706
www.rlpgbooks.com

Estover Road, Plymouth PL6 7PY, United Kingdom

Copyright © 2011 by Theodore Hailperin

All rights reserved. No part of this book may be reproduced in any form or by any
electronic or mechanical means, including information storage and retrieval systems,
without written permission from the publisher, except by a reviewer who may quote
passages in a review.

British Library Cataloguing in Publication Information Available

Library of Congress Cataloging-in-Publication Data

Library of Congress Cataloging-in-Publication Data on file under LC#2010026427
ISBN: 978-61146-010-0 (cl. : alk. paper)
eISBN: 978-161146-011-7

∞™ The paper used in this publication meets the minimum requirements of American
National Standard for Information Sciences—Permanence of Paper for Printed Library
Materials, ANSI/NISO Z39.48-1992.

Printed in the United States of America

CONTENTS

PREFACE

My interest in the connection of probability with logic was initiated by a reading of J. M. Keynes' *A Treatise on Probability*. This interest was furthered during a lengthy engagement with Boole's *Laws of Thought*, resulting in the writing of *Boole's Logic and Probability* (Hailperin 1976, 2nd ed. 1986). There is an extended historical presentation in *Hailperin* 1988 of matters pertaining to probability logic before its formalization as a logic. Subsequent publications on this topic included a book, *Sentential Probability Logic* (Hailperin 1996), containing a historical account but limited to the relation of probability with sentential logic. A number of publications then followed with further developments and applications, the present publication being an organized presentation of these further developments and includes an extension of probability logic to quantification language.

Published as separate papers over a period of time, a fair amount of editorial reorganization was needed to have them here all together as a unified subject. Nevertheless, being close enough to the original papers, I wish to express my thanks to the copyright owner-publishers for permission to so reproduce them. For papers in the Bibliography listing with dates 1988, 1997, 2006, 2007, 2008, I thank Taylor & Francis for this permission, and to Kluwer Academic Publishers for the one with date 2000. There is also some new material. Of particular note in that respect is the Main Theorem for probability logic (§3.2).

As with *Sentential Probability Logic*, Max Hailperin was a reliable resource for resolving TEX typesetting and formatting difficulties in producing the ms. He also supplied programs for producing the figures here in §§1.7, 3.4, 4.1 as well as obtaining, with use of the computer program *Mathematica*, the solution of the linear algebraic equation-inequation system in §1.7. For all this, and many helpful suggestions, I express my sincere appreciation and thanks.

Logic with a Probability Semantics

INTRODUCTION: AN OVERVIEW

1. This monograph is in large measure a continuation of *Sentential Probability Logic* (*Hailperin 1996*, hereafter to be cited as SPL). Its object is to extend the sentential probability logic there introduced, to the quantifier level and to present material making use of the extension. However the language with quantifiers we shall be using as a basis is not the customary well-known one but one that is in all respects ontologically neutral—in particular, there is no mention of individuals (arguments for predicates) or predicates. With its spare character, appealing only to the essential formal properties of quantifiers, one is able to focus exclusively on logical fundamentals. Application of this probability logic to languages that have sentences with individuals and predicates is not excluded—it just isn't required of a language that it must be of that form and, in particular as we shall see, the individuals-predicate structure of a sentence plays no role in our probability logic.

2. A large part of SPL is devoted to historical origins, to illustrative examples, to applications, and to related work of other authors. Rather than referring the reader to SPL to pick out from it those parts which are specifically needed for the present study, we thought it would be a convenience to have such material summarized here in an initial Chapter 1. At the same time this will afford us an opportunity to include some new material relating to sentential probability logic, as well as clarifying its exposition at a few places.

11

3. Probability logic as we view it is a form of logic which uses the same formal language as does verity logic but has a more general semantics. Our summarizing Chapter 1 opens with a description of the basic notions of verity sentential logic but so chosen as to foreshadow that for probability logic. Thus, instead of truth tables, verity functions are used to characterize how verity values (*true, false*) accrue to logical formulas when values are assigned to atomic sentence components. Analogously, our probability functions, a generalization of verity functions, serve to determine probability values for formulas. But now the assignment of probability values is not to atomic components but to constituents (e.g., if A_1 and A_2 are the two atomic sentences of a formula then its constituents are $A_1 A_2$, $\neg A_1 A_2$, $A_1 \neg A_2$, $\neg A_1 \neg A_2$). When properly specified (see §1.2 **II** in Chapter 1 below) such assignments are the probability models which determine a probability value for every sentence of the formal ($\neg$, $\wedge$, $\vee$)-language (supposing that $\neg$, $\wedge$, $\vee$ are its logical constants). Specific application of probability logic is made by selecting an appropriate probability model, just as for verity logic with an appropriate verity model.

It is to be emphasized that the probability logician's official business is not in the determination of what probability values are to be assigned to constituents so as to have a probability model, but rather that of the applied probabilist. She (or he) is the one who tosses the coin, draws balls from an urn, makes statistical analyses, or uses epistemic considerations, and chooses the probability model to be used. In probability logic probability values are assigned to sentences, not to (mathematical representations of) events as in mathematical probability theory.

4. The fundamental (inferential) notion of logical consequence for sentential probability logic (§1.2 **III** below) is a generalization of that for verity logic with its $V(\phi) = 1$, V a verity function, being replaced with $P(\phi) \in \alpha$, P being a probability function and α being a subset of the unit interval $[0, 1]$. Using this form as our basic semantic statement for probability logic—rather than an equality $P(\phi) = a$, a being a numerical value in $[0, 1]$—makes for a significant extension of the notion of logical consequence. For treating finite stochastic situations sentential probability logic is equally as capable as Kolmogorov probability spaces, where probability values are assigned to sets. (See §1.2 **IV** below)

5. Validity of (sentential) probability logical consequence being appropriately defined, of special interest is the result (stated in §1.3 below, carried over from SPL, §4.6) that when the subsets involved in a logical consequence statement are subintervals of $[0,1]$ with explicitly given rational end points, then there is an effective procedure for determining whether or not the statement is a valid one.

6. Development of a conditional-probability logic extending probability logic entails first of all enlarging sentential logic with a new binary connective, the suppositional. It involves having a third semantic value in addition to 0 and 1, these two still retaining their meaning of *false* and *true*. The semantics for a 2-to-3 valued logic was described in SPL §0.3, and illustrated there in §0.4 with an early application by computer scientists to logic design of computer circuits. The *suppositional*, defined as a 2-to-3 valued binary connective, was formally introduced and compared with the two-valued truth-functional conditional. As a side observation, unconnected with our probability theme, it was noted that using the suppositional in place of the truth-functional conditional as an interpretation for the " If A, then B" of ordinary discourse can resolve some puzzling instances. An example of this was given in SPL, pp. 247–248 where, in briefly mentioning this topic, about which much has been written, we were duly mindful of Callimachus' epigram, as reported by Sextus Empiricus, "Even the crows on the roofs are cawing about the nature of the conditionals."[1]

However our central interest regarding the suppositional is not in the philosophical problem of explicating "implication" as used in ordinary discourse, but that of establishing a conditional probability logic, i.e., a logic which includes the notion of conditional probability. Here the suppositional connective plays a key role. Section 1.4 below summarizes suppositional logic and §1.5 the conditional-probability logic (of sentences without quantifiers) which is based on it.

7. An analysis of an old paradox—the so-called "statistical paradox"— brings out some differences between verity and probability logical consequence (§1.6, below). Not surprisingly, in the case of probability and particularly conditional-probability inferences, inconsistencies in probability

---

[1] *Kneale and Kneale 1962* p. 128.

value assignments to premise formulas are much harder to spot, especially if schematic letters for formulas or parameters rather than actual numbers, are used. Our discussion of this example brings this out.

8. The concluding section in Chapter 1 is a discussion of the topic of combining evidence—'evidence' being taken as the condition in a statement of conditional probability. The interest is in the relation between the conditional probability of two pieces of evidence first when they are taken separately and then when the two are combined. We give a real life instance of this, using evidence that arose in a sensational murder trial. While two separate pieces of evidence individually may not be particularly strong, but when combined the logical connections between them can produce, as shown by this example, a marked change.

9. Chapter 2 presents the verity quantifier logic serving as the basis for probability quantifier logic.[2] Its formal syntactic language, however, is not that of the usual first-order predicate kind, but for an ontologically neutral logic one in which the atomic sentences have, as in sentential logic, no prescribed logical or linguistic structure. Of course, in an application these sentences may be interpreted as being composed of the familiar predicates with individuals as arguments. The sentences of formal ON logic are distinguished not by subject predicate structure but by being tagged with a variety of subscripts (to be described). While this form of a quantifier logic was not originated for the purpose of viewing probability theory denumerably—as was Borel's purpose in a 1909 paper—it turned out to be aptly suitable. (See §§3.3, 3.7, below.)

As with sentential verity logic, so also with quantifier verity logic, quantifier formulas (closed ones) acquire truth values by means of verity functions and assignments of truth values to atomic sentences. The definition of a verity function now includes means for specifying truth values for quantified formulas. A model for ontologically neutral logic is, as with sentential logic, an assignment of a verity value (0 or 1) to each atomic sentence and, again, for any such model there is a unique verity function that provides a verity value for each closed formula (Theorem 2.11). Likewise, the definition of a

---

[2]Our exposition here includes developments improving the presentation in *Hailperin 1997* and in *2000*.

valid logical consequence is the same, namely, for every verity function, the premise formulas having verity value 1 implies that the conclusion formula also has value 1.

An axiomatic formulation for ontologically neutral quantifier logic is presented, outwardly no different from that for first-order predicate logic when expressed in terms of schematic letters for formulas (§§2.2, 2.3). What makes for a difference when actual ON formulas are involved is the semantics.

The suppositional connective (§1.4) is then adjoined to this quantifier logic, its semantics in this context specified, and some properties are derived (§2.4).

10. In Chapter 3 construction of a probability logic for quantifier languages proceeds along lines developed for sentential languages. First there is a definition of a probability function for the language, now including in the list of its properties an additional one for sentences with quantifiers (P4 in §3.1). A fundamental, though somewhat difficult theorem to prove, establishes that a probability model for such a language, appropriately defined, can be uniquely extended so as to be a probability function on the language (§3.2).

A special historically based section with heading "Borel's denumerable probability" (§3.3) makes use of material from the first serious consideration of "infinite" cases in probability theory. Our probability logic for quantifier language is ideally suited for this "infinite" aspect of Borel's probability theory as the central properties and results are readily incorporated in our basic definitions, or in readily established theorems.

There is a section (§3.5) that compares the Kolmogorov spaces approach to probability theory with the logical representation here to be presented. A discussion of Bernoulli's law of large numbers is used as an example for comparison.

Since the new quantifier logic which was introduced can have not only verity but also probability as a semantic value, a general discussion of the fundamental notion of logical consequence seemed needed. This is the content of §3.6.

11. A section (§3.7) is devoted to a comparison of our quantifier prob-

ability logic and the "denumerable probability" of *Borel 1909* and to a response to the critique of Borel's ideas contained in *Barone & Novikoff 1978*.

12. Our last extension of logical language adds the suppositional to quantifier language, and defines a suitable probability function for the language. In the resulting extended probability logic we offer a resolution of the "paradox of confirmation"— its origins dating back to 1945—on which many philosophical papers have been written.

13. In concluding this Introduction we acknowledge that there are probabilists who find the Kolmogorov axiomatic set-measure-theoretic basis for probability theory so successful that it is hard for them to envision room for any other. Here, for example, is an expression of the dubiousness of ventures that have probability on sentences (*Łoś 1955*, p. 135).:

> The disadvantages of practising the calculus of probability on sentences, which discourage mathematicians from proceeding in this direction, are as follows:
>
> 1° Fields of sentences are always denumerable what implies that a Boolean algebra composed of sentences is at most denumerable. This renders it impossible to conduct the probabilistic investigation with regard to problems which require a non-denumerable field of events; in particular such a theory makes it impossible to reconstruct Lebesgue's measure on an interval (see §1).
>
> 2° An algebra formed of sentences cannot be a σ-algebra, for it is easy to show that no denumerable Boolean algebra (i.e. of the power $\aleph_0$ exactly) is a σ-algebra.
>
> 3° The fact that events are sentences has never been properly utilized in the theory of probability built on sentences (with the exception of philosophical speculations of dubious value).

With regard to Łoś's 1°, just because a theoretical approach is capable of handling problems of varying degrees of complexity doesn't mean that it has to. Thus continuum mathematics, making use of the nondenumerable real number system doesn't need to be employed for all problems using numbers—e.g., those involving just the rationals, or just the integers. Not just aesthetics but also understanding is compromised by employing stronger assumptions and methods than what are needed.

Moreover, what meaning is there to a field of a non-denumerably many *events*? One can (and many do) conceive of a *mathematical abstraction* which is a set of non-denumerably many real numbers, but of that many *events*? As will be noted in a footnote in §3.5 below, Kolmogorov's response to an objection of this nature was to declare it "unobjectionable from an empirical standpoint" since the uses of these "ideal elements", as how he refers to them, leads to the determination of the probability of actual events.

As for 2°, while the notion of σ-additivity (of probability on sets) does not directly apply to probability logic in quantifier sentences, we do have a close substitute namely 'countable additivity'. Theorem 3.43 below shows that the disjunction of any specified denumerable sequence of mutually exclusive sentences, each with a probability value, has a determined probability value.

Coming to 3°, the reference decrying those who utilize the "fact that events are sentences" does not apply to us. In our probability logic sentences may refer to, or denote, events but are not identified with them, anymore than in verity logic are sentences identified with, i.e. as being, the facts or events described.

CHAPTER 1

SENTENTIAL PROBABILITY LOGIC

§1.1. Verity logic for $\mathcal{S}(\neg, \wedge, \vee)$

We begin with a brief presentation from a semantic viewpoint of the sentential portion of verity (true, false) logic. The emphasis is on the fundamental notion of logical consequence. After stating the definition of this notion for verity logic in customary form it is then slightly modified in order to bring out that the definition of logical consequence for probability logic to be introduced in §1.2 is, as a generalization, naturally related to it. A reason why the comparison is being made on a semantic (logical consequence) basis, rather than a formal syntactic (linguistic structure) one, is given below in item **V**, end of §1.2.

The symbols of our formal sentential language $\mathcal{S}$ are standard: a potential infinite sequence of atomic sentences $A_1, \ldots, A_n$ which can be of any length, together with logical connectives $\neg$ (not), $\wedge$ (and), $\vee$ (or), in terms of which the sentences of $\mathcal{S}$ are compounded.[1] As for its semantics there are the two verity values 0 and 1. These are assumed to have the arithmetical properties of the numbers 0 and 1 so as to conveniently express the logico-semantic properties of *false* and *true*. Not only for this reason are these two numbers useful but they also will enter as semantic values for probability logic, along with the numbers in-between so that, as we shall see, sentential probability logic is a natural extension of sentential verity logic. To define the semantic properties of the connectives for verity logic

---

[1] As is well known, other choices of logical connectives can be made. These shown here are somewhat more convenient for depicting the evolvement of probability logic from verity logic.

we make use of the notion of a verity function. A function $V \colon \mathcal{S} \to \{0, 1\}$, from the sentences of $\mathcal{S}$ to the set of verity values $\{0, 1\}$, is a *verity function* on $\mathcal{S}$ if it has for any sentences ϕ and ψ of $\mathcal{S}$, the following properties:

$$V(\neg\phi) = 1 - V(\phi)$$
$$V(\phi \wedge \psi) = \min\{V(\phi), V(\psi)\}$$
$$V(\phi \vee \psi) = \max\{V(\phi), V(\psi)\}.$$

It is then easy to show that with respect to any given verity function the connectives have their usual properties as defined by truth tables, and that for a given V once values are assigned to the atomic sentences of $\mathcal{S}$, then every sentence of $\mathcal{S}$ has a uniquely defined V value. An assignment of verity values to the atomic sentences of $\mathcal{S}$ is a *verity model*. If M is a verity model we write 'V_M' for the uniquely determined associated verity function which assigns a verity value to each formula of $\mathcal{S}$. Conversely, for any verity function V defined on $\mathcal{S}$ there is a uniquely determined model M, namely that which assigns to an atomic sentence the value that the V does.[2]

For $\psi, \phi_1, \ldots, \phi_m$ that are sentences of $\mathcal{S}$, we define ψ to be a *verity logical consequence* of $\phi_1, \ldots, \phi_m$ if the following holds:

for all models M,
$$\text{if } V_M(\phi_1) = 1, \ldots, V_M(\phi_m) = 1, \text{ then } V_M(\psi) = 1. \tag{1}$$

The customary notation for this property is '$\phi_1, \ldots, \phi_m \models \psi$' or simply '$\models \psi$' if there are no premises (i.e., when ψ is truth-functionally valid). Note that the symbol '$\models$' carries the entire semantic meaning that is expressed more fully by (1). Since to each verity model there uniquely corresponds a verity function we can replace 'for all models M' by 'for all verity functions V', drop the subscript 'M' from 'V_M' and, understanding the initial universal quantifier 'for all verity functions V', reduce (1) to the simple

$$\text{if } V(\phi_1) = 1, \ldots, V(\phi_m) = 1, \text{ then } V(\psi) = 1 \tag{2}$$

---

[2]Note that $\mathcal{S}$, M and V_M are defined in terms of $A_1, \ldots, A_n$ for any $n \geq 1$, though there is no indication of this on the symbols.

or the simpler

$$\phi_1, \ldots, \phi_m \vDash \psi, \tag{3}$$

widely used to depict this notion. This definition of verity logical consequence will be of interest by way of comparison when we come to state
the one for probability logical consequence where the notion of probability
function replaces that of verity function.

The notion of logical consequence to be introduced in the next section,
where 'probability' replaces 'verity', bears little resemblance to (2). But the
existence of a relationship can be brought out by generalizing components
of the form '$V(\phi) = 1$' in (2) to '$V(\phi) \in \alpha$', where α can be any non-empty
subset of $\{0, 1\}$. Making such a change in the premises results in nothing
new since $V(\phi) \in \{1\}$ is equivalent to $V(\phi) = 1$, $V(\phi) \in \{0\}$ is equivalent
to $V(\neg\phi) = 1$ and any $V(\phi) \in \{0, 1\}$, having vacuous content, i.e., being
true for any ϕ, can be deleted. Though if $V(\psi) \in \{0, 1\}$ were to occur in
the conclusion then some information is conveyed if it were the strongest
conclusion that can be obtained, namely that the information presented in
the premises are insufficient to determine a verity value for ψ. While such
a generalization to sets of truth-values as just described has no significance
in the case of verity logical consequence, we shall later see that for the
case of probability logical consequence, a marked enlargement of the range
of application of the notion of consequence occurs when using subsets of
probability values rather than just a single value.

From this sketch of verity logical consequence we now turn to that for
our probability logical consequence which, though a natural generalization
of the one for verity logic, is more complicated.

§1.2. Probability logic for $\mathcal{S}(\neg, \wedge, \vee)$

I PROBABILITY FUNCTIONS. Although '$\vDash \phi$' is customary for expressing
validity of ϕ in formal verity logic, instead of it we shall now use '$\vDash_{tf} \phi$' to
express that ϕ is truth-functionally valid. This frees the bare '$\vDash$' for us to
use with probability logic, obviating then the need for two distinguishing

marks on '$\vDash$' since, as we shall see, both kinds of validity can be occurring though that for sentences being quite subsidiary.

For sentential probability logic the formal syntax language is $\mathcal{S}$, the same as that for verity logic. In $\mathcal{S}$ we will make use of the truth-functional conditional '$\rightarrow$' and equivalence '$\leftrightarrow$' both defined as usual in terms of $\neg$, $\wedge$, $\vee$. In place of verity functions we now have probability functions defined as follows.

Corresponding to the function V mapping sentences of $\mathcal{S}$ onto the two element set $\{0, 1\}$, here we have a function $P\colon \mathcal{S} \rightarrow [0, 1]$, from sentences of $\mathcal{S}$ to the reals of the unit closed interval $[0, 1]$, defined to be a *probability function on $\mathcal{S}$* if it has the following properties:

For any ϕ and ψ of $\mathcal{S}$,

P₁.   If $\vDash_{tf} \phi$, then $P(\phi) = 1$.

P₂.   If $\vDash_{tf} \phi \rightarrow \psi$, then $P(\phi) \leq P(\psi)$.

P₃.   If $\vDash_{tf} \phi \rightarrow \neg\psi$, then $P(\phi \vee \psi) = P(\phi) + P(\psi)$.

It then readily follows that if $\vDash_{tf} \phi \leftrightarrow \psi$, then $P(\phi) = P(\psi)$ and, more generally, where ϕ' comes from ϕ by replacing an occurrence of ρ by σ, that

$$\text{if } \vDash_{tf} \rho \leftrightarrow \sigma, \text{ then } P(\phi) = P(\phi').$$

It is readily seen (SPL, p. 190, Theorem 4.22)[3] that a probability function whose range is the two-element set $\{0, 1\}$ is a verity function, and that every verity function is a probability function. Clearly the properties of probability functions, having ranges that are subsets of $[0,1]$, are much more extensive than those of verity functions. Here is a list of some easily derivable elementary ones:

(a) $P(\phi \vee \neg\phi) = 1$

(b) $P(\phi) + P(\neg\phi) = 1$

(c) $P(\phi) = P(\phi\psi) + P(\phi\overline{\psi})$ ($\overline{\psi} := \neg\psi$, $\phi\psi := \phi \wedge \psi$)

(d) $P(\phi \vee \psi) = P(\phi) + P(\psi) - P(\phi\psi)$.

---

[3] Recall that 'SPL' stands for *Sentential Probability Logic*, i.e., *Hailperin* 1996.

There are alternative ways of expressing the defining properties of a probability function on S: sacrificing the nice parallelism in the hypotheses of P1–P3 one can replace P2 by (b) and P3 by (d) of the above list and have an equivalent definition of a probability function on S. (In the parenthetical remark accompanying (c) the symbol ':=' is an abbreviation for 'is defined as'.)

II PROBABILITY MODELS. The definition of logical consequence for probability logic makes use of probability models. However, unlike verity models which are assignments of verity values to the atomic sentences of S, in the case of *probability models* the probability values are assigned to *constituents of S* ("constituents" is what Boole, who introduced them into logic, called them). A constituent of S is a logical product of m atomic sentences, e.g. $B_1, \ldots, B_m$ in which none, some, or all are negated so that for m atomic sentences there are 2^m of such products. A compact notation for such a product is $K_{b_1 \ldots b_m}$ where b_i $(i = 1, \ldots, m)$ is either 1 or 0, it is 1 if B_i is unnegated and 0 if it is negated. In determining validity for verity logic it is immaterial whether verity values are assigned to the $B_1, \ldots, B_m$ or to the 2^m constituents on $B_1, \ldots, B_m$ since an assignment of verity values to the m $B_1, \ldots, B_m$ determines that of a unique constituent, and conversely; if B_i is assigned the value 1 then B_i is to appear as a conjunct in the constituent, but if B_i is assigned the value 0, then $\overline{B}_i$ is to appear in the conjunct. This is not the case with probability values where it only works if the components of the constituents are probabilistically independent.[4]

Let $S(K)$ denote the set of constituents of S. A *probability model* is a function $M \colon S(K) \to [0, 1]$ which for each i $(i = 1, 2, \ldots, m)$ assigns a value $k_{b_1 \ldots b_i}$ in $[0, 1]$ to $K_{b_1 \ldots b_i}$ such that

(i) $k_{b_1 \ldots b_i} \geq 0$

(ii) $k_1 + k_0 = 1$

(iii) $k_{b_1 \ldots b_i 1} + k_{b_1 \ldots b_i 0} = k_{b_1 \ldots b_i}.$

---

[4] Boole's General Method for solving "any" probability problem did assume (no justification was given) that any event could be analyzed as a logical compound of "simple" *independent* events, for which case it would be that their probability values do determine that of constituents on these events. (See SPL, §2.5, pp. 114-115.)

It then follows that for any m

$$\sum k_{b_1 \ldots b_m} = 1,$$

the sum being taken over all 2^m terms $k_{b_1 \ldots b_m}$ obtained as each b_i, $i = 1, \ldots, m$, takes on the value 0 or 1.[5]

Note that the values assigned to the $K_{b_1 \ldots b_m}$ may be, but need not necessarily be, determined by chance experiments or trials. Probability logic is accomodative to other kinds of interpretation for the notion of "probability" besides that of physical chance, for example, as an estimate of certainty[6] when there is inadequate knowledge.

Analogous to the extension of verity models for $\mathcal{S}$ to verity functions on $\mathcal{S}$ we have the result that an assignment of probability values to a specified subset of $\mathcal{S}$, i.e., those that are constituents, establishes one which satisfies P1–P3 for all formulas of the subsets of $\mathcal{S}$ (SPL, Theorem 4.41, taking its $\mathcal{S}$ to be the one here under discussion):

Any probability model M for $\mathcal{S}$ can be uniquely extended to be a probability function P_M on $\mathcal{S}$.

And, in the other direction, that one has

Any probability function P determines a probability model

is clear since properties (i) and (ii) of the definition of a probability model obviously hold and (iii) does by item (c) above with ϕ and ψ appropriately specified.

III PROBABILITY VALIDITY. The definition of logical consequence for sentential probability logic which we now state, is a generalization of that for verity logic (that of §1.1) in which probability models replace verity models, and subsets of $[0, 1]$ replace subsets of $\{0, 1\}$.

Let $\phi_1, \ldots, \phi_m, \psi$ be sentences of $\mathcal{S}$. We say ψ is *a probability logical consequence of* the m sentences $\phi_1, \ldots, \phi_m$ (with respect to non-empty subsets

---

[5]As a help in keeping matters clear note that a b_i is a numeral 0 or 1, that $b_1 \cdots b_i$ is a numeral in binary notation, that $k_{b_1 \ldots b_i}$ a real number in [0,1], and $K_{b_1 \ldots b_i}$ a sentence in $\mathcal{S}(K)$.

[6]An idea going back at least to J. Bernoulli. See e.g., SPL §1.2.

$\alpha_1, \ldots, \alpha_m, \beta$ of $[0, 1]$) if the following condition is satisfied:

For any probability model M(with probability function P_M) :

$$\text{if } P_M(\phi_1) \in \alpha_1, \ldots, P_M(\phi_m) \in \alpha_m, \text{ then } P_M(\psi) \in \beta. \qquad (1)$$

Since to each probability model M there is a unique probability function P_M, and to each probability function P a unique model M, we can reformulate (1) to:

For any probability function P :

$$\text{if } P(\phi_1) \in \alpha_1, \ldots, P(\phi_m) \in \alpha_m, \text{ then } P(\psi) \in \beta. \qquad (2)$$

Analogously to the definition of logical consequence in verity logic we drop the universal quantifier, i.e., take P in the generality sense, and abbreviate definition (2) to

$$P(\phi_1) \in \alpha_1, \ldots, P(\phi_m) \in \alpha_m \vDash P(\psi) \in \beta. \qquad (3)$$

The clauses '$P(\phi_i) \in \alpha_i$' are its *premises* and '$P(\psi) \in \beta$' is its *conclusion*, and to make this notion precise, we assume that β is the smallest set for which (2) is the case. When there are no premises we shall refer to the conclusion as a *probability logic β-validity*.

Customarily formulas derived from the P1–P3, stated above at the beginning of this section, are what are usually considered to be 'laws' of probability, rather than validities of a logic as we have been presenting them.

The following *notational convention* is intended to bring out that a formula derived from the semantic properties P1–P3 is equivalently representable as a probability logical validity.

Let $\mathcal{F}(P(\psi_1), \ldots, P(\psi_m))$ be a formula derivable from P1–P3. Introducing the set

$$\beta_1 = \min_{x \in [0,\, 1]} \mathcal{F}(x, P(\psi_2), \ldots, P(\psi_m))$$

we have that $\mathcal{F}(P(\psi_1), \ldots, P(\psi_m))$ being derivable is equivalent to the β_1-validity statement

$$\vDash P(\psi_1) \in \beta_1.$$

Similarly, with β_i having x in the ith argument position in place of $P(\psi_i)$ we have for any i, $1 \leq i \leq m$

$$\vDash P(\psi_i) \in \beta_i.$$

All of these m statements of validity express the same fact that the formula $\mathcal{F}(P(\psi_1), \ldots, P(\psi_m))$ is a consequence of P1–P3. In place of these m validity statements we can convey the same information just by writing

$$\vDash \mathcal{F}(P(\psi_1), \ldots, P(\psi_m))$$

with the adoption of this meaning for '$\vDash$' when not accompanied by a statement of membership in a set, so bringing the logical notation in conformity with the 'mathematical' one in general used by 'mathematical' probabilists.

Thus each of the equations in the list (a)–(d) can have a '$\vDash$' prefixed to it. Some statements derivable from P1–P3 may not be equations so that, for example, we also have the validities

(e) $\vDash \max\{P(\phi), P(\psi)\} \leq P(\phi \vee \psi)$

(f) $\vDash P(\phi \wedge \psi) \leq \min\{P(\phi), P(\psi)\}.$

IV KOLMOGOROV SPACES. The following result (SPL, Theorem 4.71) is to the effect that, as far as finite stochastic situations are concerned, there is no loss of generality in using probability models of probability logic instead of Kolmogorov probability spaces:

There is an effective pairing of probability models with finite probability spaces (of sets) which is sense-preserving, i.e., is such that the corresponding members in a pairing model the same stochastic situation.

V SYNTACTIC FORMULATION? One may wonder: Although verity logic has a purely syntactic formulation why isn't there one for probability logic? Indeed, having a semantic one for sentential verity logic didn't come into existence until the early part of the twentieth century. What makes for the difference, i.e., the preference for a syntactic version, is the extreme simplicity of the verity semantics so that it could be taken over by the syntax. To see this we describe an example of the transformation of a verity logic assertion from a semantic to a syntactic form as could occur in a verity logic.

In the first place one can limit semantic statements to those of the form $V(\phi) = 1$ since, 0 and 1 being the only semantic values, a statement of the form $V(\phi) = 0$ is equivalent to $V(\neg\phi) = 1$. Moreover, a set of m premises $V(\phi_1) = 1, \ldots, V(\phi_m) = 1$ is equivalent to a single premise $V(\phi_1\phi_2 \cdots \phi_m) = 1$. Additionally, a consequence relation $V(\phi) = 1 \vDash V(\psi) = 1$ is equivalent to one with no premises, i.e., to $\vDash V(\phi \rightarrow \psi) = 1$. Thus one can have, for an arbitrary V, a sequence of statements each one conveying equivalent information:

$$V(\phi_1) = 1, \ldots, V(\phi_m) = 1 \vDash V(\psi) = 1 \tag{4}$$

$$V(\phi_1\phi_2 \cdots \phi_m) = 1 \vDash V(\psi) = 1 \tag{5}$$

$$\vDash V(\phi_1\phi_2 \cdots \phi_m \rightarrow \psi) = 1 \tag{6}$$

$$\vdash \phi_1\phi_2 \cdots \phi_m \rightarrow \psi, \tag{7}$$

the first three being semantic in form, the last syntactic. The introduction of a semantic form and its equivalence with the syntactic one was not recognized and shown to be equivalent to the syntactic one until the 1920's. (For a historical account see, e.g. *Kneale and Kneale* 1962, Chapter XII.)

Since truth-functional validity and deducibility from axioms for sentential logic are effectively equivalent one could have in P1-P3 of the definition of a probability function either the semantic '$\vDash_{tf}$', or the syntactic '$\vdash$' denoting provability from axioms for sentential logic.

§1.3. Interval-based probability logic

As observed earlier the use of subsets of verity values (i.e., $\{0\}, \{1\}, \{0,1\}$) in sentential logic is of no particular interest. This is not the case for subsets of probability values. When in a statement of probability logical consequence

$$P(\phi_1) \in \alpha_1, \ldots, P(\phi_m) \in \alpha_m \vDash P(\psi) \in \beta,$$

the sets α_i ($i = 1, \ldots, m$) are sub*intervals* of $[0, 1]$ the problem of determining the optimal β, i.e., the narrowest interval β for which the statement holds, is convertible to a problem in linear algebra (SPL, §4.5). To be able to make this conversion the ϕ_i and ψ have to be explicitly given sentences of $\mathcal{S}$ since the method requires that these sentences be expressed as logical sums of constituents in terms of atomic sentences. We illustrate this method using a simple example. Note that $P(\phi) \in [a, b]$ is equivalent to the double inequality $a \leq P(\phi) \leq b$, and that $P(\phi) \in [a, a]$ is equivalent to $P(\phi) = a$.

As our example we choose the probabilistic form of *modus ponens*:

$$P(A_1) = p, \ P(A_1 \rightarrow A_2) = q \vDash P(A_2) \in \beta. \tag{1}$$

Clearly having $\beta = [0, 1]$ is too weak to be of interest. What is wanted for β is the smallest set, expressed in terms of the parameters p and q, which is contained in every β making (1) a valid logical probability consequence. Conversion of the problem to algebraic form is accomplished as follows. Replace A_1 by $A_1 A_2 \vee A_1 \overline{A}_2$ and $A_1 \rightarrow A_2$ by $A_1 A_2 \vee \overline{A}_1 A_2 \vee \overline{A}_1 \overline{A}_2$. Then making use of P being a probability function distribute it over the disjuncts. Letting k_1, k_2, k_3, k_4 stand for the respective probabilities $P(A_1 A_2)$, $P(A_1 \overline{A}_2)$, $P(\overline{A}_1 A_2)$, $P(\overline{A}_1 \overline{A}_2)$, the pair of premise conditions in (1) is converted to the linear equation-inequation system:

$$
\begin{aligned}
k_1 + k_2 &\qquad\qquad = p \\
k_1 &\quad + k_3 + k_4 = q \\
k_1 + k_2 &+ k_3 + k_4 = 1 \\
k_1, k_2, &\ k_3, k_4 \geq 0.
\end{aligned}
\tag{2}
$$

The interest here is in the set of values that the variable $w = P(A_2) = k_1 + k_3$ can take on as the k_1, k_2, k_3, k_4 range over the reals, subject to conditions (2). Using simple algebraic properties (for the details see §4.5 of SPL) it turns out that, with the probability conditions

$$p \leq 1, \ q \leq 1, \ 1 \leq p + q$$

on the parameters p, q, the smallest set of values for $P(A_2)$ is the interval $[p + q - 1, \ q]$. (For explanation on the ideas behind the technique of finding this interval see SPL, pp. 199-201.) Another example of this notion is the probabilistic form of the " hypothetical syllogism" (SPL, Theorem 4.53):

$$P(A_1 \to A_2) = p, \ P(A_2 \to A_3) = q \vDash P(A_1 \to A_3) \in [p + q - 1, 1].$$

The following examples (SPL, Theorem 4.545) feature the two inferences just cited when their two premises are each near 1, the nearness being expressed in terms of the 'uncertainty' ϵ, the difference from 1:

(a) $P(A_1) \in [1 - \epsilon, 1], \ P(A_1 \to A_2) \in [1 - \epsilon, 1]$

$$\vDash P(A_2) \in [1 - 2\epsilon, 1]$$

(b) $P(A_1 \to A_2) \in [1 - \epsilon, 1], \ P(A_2 \to A_3) \in [1 - \epsilon, 1]$

$$\vDash P(A_1 \to A_3) \in [1 - 2\epsilon, 1].$$

In each of these examples the sentence in the conclusion is a (verity) necessary consequence of those in the premises. This need not be the case, the method still being applicable. For example

$$P(A_1 \to A_2) = p \vDash P(A_2 \to A_1) \in [1 - p, 1]$$

is a valid probability logical consequence relating the probability of a conditional with that of its converse.

It can be shown (SPL, §4.6) that this linear algebra method provides a decision procedure for an explicitly given probability logical consequence relation. By a probability logical consequence relation being *explicitly given* we mean that the sentences $\phi_1, \ldots, \phi_m, \psi$ are expressed in terms of atomic sentences and that the end-points of the intervals involved in the premises are given *rational* numbers. We may express this result (SPL, Theorems 4.62, 4.63) as follows:

For any explicitly given relation of the form

$$P(\phi_1) \in [a_1, b_1], \ \ldots, \ P(\phi_m) \in [a_m, b_m] \vDash P(\psi) \in [l, u],$$

the a_i, b_i, $(i = 1, \ldots, m)$ being rationals, there is an effective procedure for determining whether it is a valid sentential probability logical consequence relation with $[l, u]$ the optimal interval for $P(\psi)$.

§1.4. Sentential suppositional logic

This section describes the modified sentential logic that supports conditional-probability logic (to be presented in the next section), corresponding to the way that verity logic supports (unconditional) probability logic. The material is condensed from SPL, §§0.3, 0.4 and 5.6.

Suppositional logic has, in addition to 0 and 1, a third semantic value, u, with the intuitive meaning "either 0 or 1, it is undetermined, unknown, or of no interest which".[7] Guided by this meaning we ascribe to u the following "arithmetical" properties so as to conveniently characterize its properties with respect to the logical connectives:

$$0 \le u \le 1$$
$$1 - u = u$$
$$\min(0, u) = 0, \quad \min(1, u) = u$$
$$\max(0, u) = u, \quad \max(1, u) = 1$$
$$\min(u, u) = \max(u, u) = u,$$

min and max being functions symmetric in their arguments.[8]

The (formal) syntax language for (sentential) suppositional logic is that of verity logic supplemented with an additional binary connective symbol '$\dashv$' called the *suppositional* ('$B \dashv A$' being read either as 'B, supposing A' or 'supposing A, then B'). The set of sentences of the logic is denoted by '$\mathcal{S}^u$'; it has $\mathcal{S}$ as a proper subset. In place of verity functions here we have *suppositional functions*. These are functions $U \colon \mathcal{S}^u \to \{0, u, 1\}$, i.e., from sentences of $\mathcal{S}^u$ to the set $\{0, u, 1\}$ having, for sentences Φ, Ψ of $\mathcal{S}^u$, the following properties:

$$U(\neg\Phi) = 1 - U(\Phi)$$
$$U(\Phi \wedge \Psi) = \min\{U(\Phi), U(\Psi)\}$$
$$U(\Phi \vee \Psi) = \max\{U(\Phi), U(\Psi)\} \qquad (1)$$
$$U(\Psi \dashv \Phi) = \begin{cases} U(\Psi) & \text{if } U(\Phi) = 1 \\ u, & \text{otherwise.} \end{cases}$$

---

[7]For further discussion of this notion see SPL, §§0.2–0.4, 5.6.

[8]These "arithmetic" properties of u are obtained on the basis of assigning to it the meaning of the indefinite description $\nu x(x \in \{0, 1\})$. For details see SPL, §§0.2, 0.3.

Continuing with the exposition, it is specified that *the verity values 0 and 1 are the only ones assignable to atomic sentences* and that the value u enters only semantically with the use of the connective ' $\dashv$ ' as shown in the last line of display (1). Then, since only 0 and 1 being assignable to atomic sentences, a basic suppositional 'truth' table with a heading of n atomic sentence symbols has under it 2^n rows of possible truth value assignments, and alongside a value column that, in addition to 0's and 1's, may also contain occurrences of u. Thus sentential suppositional logic is a 2-to-3 valued logic.

Carrying over from sentential verity logic its definition of a formula being valid—that its value is 1 under all possible assignments of verity values to its atomic components—is not suitable for suppositional logic. For instance, if ϕ is a sentence in $\mathcal{S}$ having any number of both 0 and 1 as values then 'ϕ, supposing that ϕ' ought to be (suppositionally) valid. But its semantic table (condensed[9]) is

$$
\begin{array}{c|c}
\phi & \phi \dashv \phi \\
\hline
1 & 1 \\
0 & u
\end{array}
$$

which doesn't assign to $\phi \dashv \phi$ all 1 values. Rather than all 1's, the adopted requirement for validity in suppositional logic is that there be no 0 values *and* at least one 1 value. (This last proviso serves to exclude the case of all the values being u.) Accordingly, we define a *model for suppositional logic* to be an assignment of verity values to the atomic sentences of $\mathcal{S}^u$ (which are also the atomic sentences of $\mathcal{S}$). A formula of $\mathcal{S}^u$ is then *u-valid* if the value column of its semantic table has at least one occurrence of 1 and no occurrences of 0. An inference form is *u-valid* if it preserves u-validity. Clearly the notion of a formula of $\mathcal{S}^u$ being u-valid is an effective one, i.e., one can determine in a finite number of steps whether or not ϕ is u-valid.

Theorem 5.63 in SPL lists a number of u-valid inference forms which are the same as for verity logic but with '$\dashv$' replacing '$\rightarrow$'. There are, of course, differences. Contraposition, for example, fails and likewise monotonicity, e.g., although $A_1 \dashv A_1$ is u-valid $A_1 \dashv A_1 \overline{A_1}$ is not. Despite the differences

---

[9]That is, each table entry line shown could be repeated depending on the number of ϕ's atomic components.

the relationship between the verity conditional and the suppositional is close enough to warrant our referring to ϕ as the *(suppositional) antecedent* and to ψ as the *(suppositional) consequent* in a formula of the form $\psi \dashv \phi$. (We have chosen to symbolize the suppositional with the 'Consequent' coming before the 'Antecedent' so that its probability—which we shall identify with conditional probability (in probability logic)—will look like what has been used for more than a century in mathematical probability literature, namely, $P(C \mid A)$.)

There is an important normal form for sentences of $\mathcal{S}^u$. Every sentence of $\mathcal{S}^u$ is constructed from atomic sentences and the connectives. If Φ is a sentence whose atomic sentences are $A_1, \ldots, A_n$ a semantic table for Φ has, under these as headings, 2^n rows representing all possible assignments of 0 or 1 to $A_1, \ldots, A_n$. To each such assignment one can correlate in customary fashion, one of the constituents $K_1, \ldots, K_{2^n}$. At the end of each row of the table, under the heading Φ there is a value for Φ, either 0, u, or 1. Then the formula

$$K_{i_1} \vee \cdots \vee K_{i_r} \dashv K_{j_1} \vee \cdots \vee K_{j_s}, \tag{2}$$

with only one occurrence of '$\dashv$', in which $i_1, \ldots, i_r$ are the rows in which Φ's value is 1 and $j_1, \ldots, j_s$ are the rows in which Φ's value is either 0 or 1, has exactly the same semantic table as that of Φ.[10] Two sentences of $\mathcal{S}^u$ having the same semantic table or, equivalently, the same normal form (2) (allowing for the introduction of "vacuously" occurring atomic sentences to make their set of atomic sentences the same, if needed) we call *u-equivalent*.

An important consequence of this result, that each sentence of $\mathcal{S}^u$ has a unique *suppositional normal form*, is that every sentence of $\mathcal{S}^u$ has a u-equivalent form in which there is at most one occurrence of '$\dashv$' (SPL, end of §5.6). A pictorial representation of an $\mathcal{S}^u$ sentence can be had by obtaining its normal form and then in a Venn diagram for it excising those regions corresponding to constituents not having 0 or 1 as its value (i.e., those, if any, having the value u) and shading those having the value 0. In such a picture the fundamental regions (constituents) not deleted are those in the

---

[10]If r or s is 0 (meaning that there are no constituents) then in (2) replace the respective indicated logical "sum" by $A_1\overline{A}_1$. We take the opportunity here to correct a typographical error on p. 255 of SPL: on line 14 remove the long overline from the first quantity equated to A and place it over the second one.

antecedent of (2)—the formula's universe of discourse—while those in the consequent correspond to the unshaded portion.

The letter u occurring in 'u-valid' and 'u-equivalent' emphasizes that these semantic notions are specific to suppositional logic and need not have the same properties as 'valid' and 'equivalent' in verity logic. While it is the case that u-equivalence functions like equivalence, e.g., replacement of a formula part by one that is u-equivalent to the original. But u-valid is "weaker" than valid; e.g., the formulas $A \dashv A$ and $\neg A \dashv \neg A$ are both u-valid, yet they are not u-equivalent since if one of them has the value 1 in a model the other has the value u, and vice versa.

The following is a list of some u-equivalences taken from SPL §5.6. The symbol '$\equiv_u$' expresses u-equivalence.[11]

   (a) $\phi \equiv_u \phi \dashv (A_1 \vee \overline{A}_1)$

   (b) $\neg(\psi \dashv \phi) \equiv_u \overline{\psi} \dashv \phi$

   (c) $(\psi_1 \dashv \phi_1) \wedge (\psi_2 \dashv \phi_2) \equiv_u \psi_1\psi_2 \dashv (\phi_1\phi_2 \vee \phi_1\overline{\psi}_1 \vee \phi_2\overline{\psi}_2)$

   (d) $(\psi_1 \dashv \phi_1) \vee (\psi_2 \dashv \phi_2) \equiv_u (\psi_1 \vee \psi_2) \dashv (\phi_1\phi_2 \vee \phi_1\psi_1 \vee \phi_2\psi_2)$.

The following are u-equivalences between a formula (on the left of the '$\equiv_u$') containing an occurrence of '$\dashv$' within the scope of another '$\dashv$' and one (on the right) that doesn't:

$$(e) \quad (\psi_1 \dashv \phi_1) \dashv (\psi_2 \dashv \phi_2) \equiv_u \psi_1 \dashv \phi_1\phi_2\psi_2$$

and some of its special cases,

   (f) $(\psi \dashv \phi) \dashv \sigma \equiv_u \psi \dashv \phi\sigma$

   (g) $\psi \dashv (\phi \dashv \sigma) \equiv_u \psi \dashv \phi\sigma$

   (h) $(\psi \dashv \phi) \dashv (\psi \dashv \phi) \equiv_u \psi \dashv \psi\phi$.

Note that in each of the (a)–(h) u-equivalences the sentence to the right of '$\equiv_u$' has just one occurrence of '$\dashv$'.

---

[11]Since '$\equiv_u$' is a semantic symbol and not a connective of suppositional logic to be precise we should write '$\phi \equiv_u \psi$' rather than $\phi \equiv_u \psi$, however the subscript 'u' will suffice to indicate its having a special semantic meaning.

Listing (b) along with special cases of (c) and (d) gives

(i) $\neg(\psi \dashv \phi) \equiv_u \overline{\psi} \dashv \phi$

(j) $(\psi_1 \dashv \phi) \wedge (\psi_2 \dashv \phi) \equiv_u \psi_1\psi_2 \dashv \phi$

(k) $(\psi_1 \dashv \phi) \vee (\psi_2 \dashv \phi) \equiv_u (\psi_1 \vee \psi_2) \dashv \phi$

which shows the close relationship of the suppositional with fixed antecedent, here ϕ, and the sentential logic of $\neg$, $\wedge$, $\vee$.

Some other interesting u-equivalences are[12]

(l) $\psi \dashv \phi \equiv_u \phi\psi \vee \neg\phi\mathbf{u}$

(m) $\phi(\psi \dashv \phi) \equiv_u \phi\psi$

(n) $\sigma(\psi \dashv \phi) \dashv \phi\psi \equiv_u \sigma \dashv \phi\psi.$

A note of caution: u-validity of a formula is not in general maintained under substitution for an atomic sentence. For example, $A_1 \dashv A_1$ is u-valid but not on replacing A_1 by, say, $A_2\overline{A}_2$. However, replacement of a formula (part) by a u-equivalent one does preserve u-validity.

In developing our sentential suppositional logic we were unaware that a slightly less general form had been presented earlier by D. Dubois and H. Prade (See §§3, 5 in their essay "Conditional Objects, Possibility Theory and Default Rules", and references there given, in *Crocco, Fariñas del Cerro and Herzig*, 1995.) We say slightly less general in that in their version only formulas of the form '$\psi \dashv \phi$' (to use our notation), referred to by them as a 'conditional object', were considered where ψ and ϕ can take on only the truth values 0 and 1. As mentioned above, we have shown in SPL, p. 253, that formulas allowing suppositionals within the scope of suppositionals can be reduced to u-equivalent ones of the '$\psi \dashv \phi$' form with just one occurrence of '$\dashv$'. Instead of u-equivalence the basic Dubois-Prade semantic relation is "semantic entailment"—a formula Φ semantically entails Ψ if under no semantic assignments does the value of Φ exceed that of Ψ (assuming the ordering $0 \leq u \leq 1$). It is clear that two formulas that semantically entail each other are u-equivalent. If we call their notion "u-entails" (parallel to our "u-equivalent") and symbolize it by '$\Rightarrow_u$' then, as examples, the

---

[12]The $\mathbf{u}$ in (l) is a logical constant whose value is u in any model.

following is a list of u-entailments taken from their essay (p. 313). They are all readily checked, as the ϕ, ψ, and χ can only take on 0 or 1 as values.

$$\text{(o)} \quad (\psi \dashv \phi) \wedge (\chi \dashv \phi) \Rightarrow_u \chi \dashv \phi\psi$$

$$\text{(p)} \quad (\psi \dashv \phi) \wedge (\chi \dashv \phi\psi) \Rightarrow_u \chi \dashv \phi$$

$$\text{(q)} \quad (\psi \dashv \phi) \wedge (\chi \dashv \phi) \Rightarrow_u \psi\chi \dashv \phi$$

$$\text{(r)} \quad (\psi \dashv \phi) \wedge (\phi \dashv \psi) \wedge (\chi \dashv \psi) \Rightarrow_u \chi \dashv \phi$$

$$\text{(s)} \quad \chi \dashv (\phi\psi) \Rightarrow_u (\neg\phi \vee \chi) \dashv \psi$$

$$\text{(t)} \quad (\chi \dashv \phi) \wedge (\chi \dashv \psi) \Rightarrow_u \chi \dashv (\phi \vee \psi).$$

Our interest here is making use of the u-value in conjunction with probability values so as to have a conditional probabiity logic—to which we now turn.

§1.5. Conditional-probability logic

Scattered throughout §§2.7, 3.1, 5.8 of SPL there are historical remarks on the idea of a "conditional event" associated with conditional probability. Preceded by preliminary discussion and development, a definition of such a notion and its associated logic was stated there in §5.7. The following is a summary presentation of this material to serve as a basis for its extension to quantifier probability logic. Additionally we will introduce in §1.6 a clarifying example solving a long standing puzzle, and in §1.7 some new results involving conditional probability logic involved in the combining of evidence.

As described in §1.2 above, probability logic generalizes verity logic by having, instead of verity value assignments to atomic sentences of $\mathcal{S}$, probability value assignments to constituents on atomic sentences. From such assignments probability values then accrue to sentences of $\mathcal{S}$ when these are expressed as logical sums of constituents. Similarly, for conditional-probability logic initial assignments are also made to constituents, which leads to probability values, but only for those sentences in the subset $\mathcal{S}$ of

$\mathcal{S}^u$. The other sentences of $\mathcal{S}^u$ are those containing one or more occurrences of '⊣'. The definition of conditional-probability to be introduced makes use of the result (SPL, p. 253 end of §5.6, restated here in the preceding section) that any such sentence can be reduced to a u-equivalent one with its suppositional normal form containing at most one occurrence of '⊣'. The conditional-probability of any sentence with occurrences of '⊣' is then defined in terms of its u-equivalent one with a single occurrence of '⊣'. The function on $\mathcal{S}^u$ which results, called a *conditional-probability function*, is denoted by P^*. (The 'P' portion of this notation serving also to represent the probability function on $\mathcal{S}$ which is part of the definition of P^*.) The range of P^* will be the interval $[0,1]$ with an additional "value" c. Just as u is an indefinite value in the set $\{0,1\}$, so c is an indefinite "numerical value" in the unit interval $[0,1]$.[13] We then have

DEFINITION OF CONDITIONAL-PROBABILITY'S P^*

Let P be a probability function on $\mathcal{S}$ and Φ a sentence of $\mathcal{S}^u$.

(i) If Φ's suppositional normal form is the same as that of a ϕ in $\mathcal{S}$ (i.e., one that has no occurrence of ' ⊣ ') then set

$$P^*(\Phi) = P(\phi).$$

(ii) If Φ's suppositional normal form is $\psi \dashv \phi$ (ψ, ϕ in $\mathcal{S}$) then set

$$P^*(\Phi) = \begin{cases} P(\psi\phi)/P(\phi) & \text{if } P(\phi) \neq 0 \\ c, & \text{otherwise,} \end{cases}$$

c being an indefinite element of $[0,1]$.

It is evident from this definition that $P^*(\Phi) = P^*(\Psi)$ if Φ and Ψ are u-equivalent (since they have the same suppositional normal form), and that $P^*(\psi \dashv \phi)$ is unchanged in value if ψ, or ϕ, is replaced by a logically equivalent sentence of $\mathcal{S}$. Moreover, given a P defined on $\mathcal{S}$ then P^* provides a value (in $[0,1] \cup \{c\}$) for each closed formula of $\mathcal{S}^u$. However P^* is not a probability function as defined in §1.2 except for limited classes of closed

---

[13]Similar to the constant of integration C in Integral Calculus which is an indefinite value in $(-\infty, \infty)$. For a discussion of the use of indefinite descriptions in logic see SPL §§0.2, 0.3.

formulas, e.g., those having the same antecedent with non-zero P value. (See (5)–(7) below and compare with (a), (b), (d), in §1.2.)

Completion of the specification of a logic for conditional-probability requires a definition of logical consequence. The definition is patterned after that for (unconditional) probability logic:

We say Ψ is a *(conditional-probability) logical consequence* of $\Phi_1, \ldots, \Phi_m$ with respect to the non-empty subsets $\alpha_1, \ldots, \alpha_m, \beta$ of $[0, 1] \cup \{c\}$, when for each probability function P on $\mathcal{S}$

$$\text{if } P^*(\Phi_1) \in \alpha_1, \ \ldots, \ P^*(\Phi_m) \in \alpha_m, \text{ then } P^*(\Psi) \in \beta.$$

Making use of the abbreviating notation introduced in §1.2, we express this by

$$P^*(\Phi_1) \in \alpha_1, \ \ldots, \ P^*(\Phi_m) \in \alpha_m \ \vDash \ P^*(\Psi) \in \beta.$$

Since P^* coincides with P when its argument is in $\mathcal{S}$, and only when its argument contains an occurrence of '⊣' does P^* differ from P, it will simplify notation if we drop the asterisk and recognize this P as being P^* by the presence of '⊣' in its argument. Additionally, having served its purpose in enabling us to introduce the notion of a conditional event, the symbol '⊣' is now to be replaced with the widely used '$|$', keeping in mind when it occurs in, or as part of an argument of P, that for us it is not a shorthand notation without independent meaning but a connective of suppositional logic. These two changes are made simply to have our notation look like what is widely used in probability literature. Occasionally, for clarity or emphasis or when '⊣' is not explicit, we shall revert to using P^* instead of P.

Just as with u being endowed with "arithmetical" properties in suppositional logic, so with c in conditional-probability logic we endow it with "real number" properties, namely having the properties $0 \leq c \leq 1$ and $0 = c \cdot 0$. Then for *any* ψ, ϕ in $\mathcal{S}$,

$$0 \leq P(\psi \mid \phi) \leq 1, \text{ and} \tag{1}$$

$$P(\psi\phi) = P(\psi \mid \phi)P(\phi), \tag{2}$$

making things look more like what one sees in mathematical probability writings.

In connection with (1) we point out that commensurability of conditional-probabilities (i.e., of any pair, one is $\leq$ the other) does not hold if one of the pair has the value c—except, as (1) shows, if the other of the pair has value 0, or 1.

Adjoining these properties, i.e., (1) and (2), to those of a probability function (P_1–P_3 of §1.2), together with those of P^* given in its definition and the u-equivalences of suppositional logic (any two being intersubstitutable as arguments of P^*), and using the notational convention of §1.2 regarding '$\vDash$' then enables the assertion of algebraic relations as logical consequences just the same as for unconditional probability. The above cited (1) and (2) can thus have a '$\vDash$' prefixed to them. Here are some additional ones (SPL, Theorem 5.30 (b), (d)):

$$\vDash P(\phi\psi\chi) = P(\phi \mid \psi\chi)P(\psi \mid \chi)P(\chi) \tag{3}$$

$$\vDash P(\sigma)P(\phi \mid \sigma) = P(\sigma(\sigma \to \phi)). \tag{4}$$

Note, in (4), the interrelationship of the suppositional '$\phi\mid\sigma$' and the conditional '$\sigma \to \phi$' in this probability context.

Some such consequence relations need premises (SPL, Theorem 5.32):

$$P(\sigma) \neq 0 \ \vDash P(\sigma \mid \sigma) = 1 \tag{5}$$

$$P(\sigma) \neq 0 \ \vDash P(\phi \mid \sigma) + P(\neg\phi \mid \sigma) = 1 \tag{6}$$

$$P(\sigma) \neq 0 \ \vDash P((\phi \vee \psi) \mid \sigma) = P(\phi \mid \sigma) + P(\psi \mid \sigma) - P(\phi\psi \mid \sigma). \tag{7}$$

(Although the hypothesis here, $P(\sigma) \neq 0$, doesn't have the $P(\sigma) \in \alpha$ form it is equivalent to it on taking α to be the (half open) subset $(0, 1]$ of $[0, 1]$.)

Significantly different from the conditional probability defined simply as a quotient of ordinary (unconditional) probabilities, we can state consequences involving additional occurences of '$\mid$' within logical connectives (SPL, Theorem 5.71):

$$\vDash P(\neg(\psi \mid \phi)) = P(\overline{\psi} \mid \phi) \tag{8}$$

$$\vDash P((\psi_1 \mid \phi_1) \wedge (\psi_2 \mid \phi_2)) = P(\psi_1\psi_2 \mid \phi_1\phi_2 \vee \phi_1\overline{\psi}_1 \vee \phi_2\overline{\psi}_2) \tag{9}$$

$$\vDash P((\psi_1 \mid \phi_1) \vee (\psi_2 \mid \phi_2)) = P(\psi_1 \vee \psi_2 \mid \phi_1\phi_2 \vee \phi_1\psi_1 \vee \phi_2\psi_2). \tag{10}$$

Here are some in which the suppositional occurs within the scope of a suppositional (SPL, Theorem 5.72):

$$\vDash P((\psi \mid \phi) \mid \sigma) = P(\psi \mid \phi\sigma) \tag{11}$$

$$\vDash P(\psi \mid (\phi \mid \sigma)) = P(\psi \mid \phi\sigma) \tag{12}$$

$$\vDash P((\psi_1 \mid \phi_1) \mid (\psi_2 \mid \phi_2)) = P(\psi_1 \mid \phi_1\phi_2\psi_2), \tag{13}$$

all this making for a much richer theory.

The following results will be used in §4.2.

THEOREM 1.51. *If ρ and σ are sentences in $\mathcal{S}$, and Φ is one in $\mathcal{S}^u$, then*

$$P^*((\Phi(\rho \mid \sigma)) \mid \rho\sigma) = P^*(\Phi \mid \rho\sigma).$$

PROOF. This follows immediately from the u-equivalence

$$(\Phi(\rho \dashv \sigma)) \dashv \rho\sigma \equiv_u \Phi \dashv \rho\sigma$$

whose proof is: if $U(\rho\sigma) = 0$ then both sides of the u-equivalence have value u, while if $U(\rho\sigma) = 1$, then they both evaluate to $U(\Phi)$. $\square$

Similarly, since $\Phi \dashv \sigma \equiv_u \Phi\sigma \dashv \sigma$, one has

THEOREM 1.52. $P^*(\Phi \mid \sigma) = P^*(\Phi\sigma \mid \sigma)$.

A SIDE REMARK. Although 'suppositional' was the name I chose for the notion symbolized by '$\dashv$' on introducing it into sentential logic and then also into conditional probability logic, I thought at that time this symbol would not be welcomed as a replacement for the '$\mid$' symbol in use in mathematical probability literature for over a hundred years.

To my surprise I recently discovered that Bayes had used the word 'supposition' in his celebrated paper (posthumously published in 1763 by Rev. Richard Price) in which he introduced what everyone now refers to as 'conditional probability.' (For a modern reprint of Bayes' paper see, e.g., *Dale* 2003, pp. 269-297.) I quote from *Dale* 2003, p. 274: "Wherefore the ratio of P to N is compounded of the ratio of a to N and that of b to N, i.e., the probability that two subsequent events will both happen is compounded of the probability of the 1st and the probability of the 2nd on supposition [sic] the 1st happens."

So, Bayes wrote "on supposition" not "on condition". Should one prefer to say "suppositional probability" or is it too late to change?

§1.6. Logical consequence for probability logic

In the definition of logical consequence for sentential probability logic (§1.2 **III** above) the basic semantic components are statements of the form $P(\phi) \in \alpha$, where P is an arbitrary probability function (becoming specific with a choice of probability model), and α is a subset of $[0, 1]$. Such subsets need not be explicitly given but may only be described—for example, the set $\{x \in [0, 1] \mid \mathcal{F}(x, P(\psi_1), \ldots, P(\psi_m))\}$, occurring in the discussion at the end of §1.2 **III**. This enlargement of the probability semantic language beyond the simple $V(\phi) = 1$ of verity logic brings in the need for sharper distinction between syntax and semantics, a distinction barely noticeable in verity logic. Indeed for verity logic one finds, rather, that the syntactic *deduction from premises* idea,

$$\phi_1, \ldots, \phi_m \vdash \psi,$$

(as in *modus ponens*: $\phi, \ \phi \rightarrow \psi \vdash \psi$) is used, being handier than the semantic

$$\text{for all } V, \quad V(\phi_1) = 1, \ldots, V(\phi_m) = 1 \vDash V(\psi) = 1.$$

For verity logic confusion of ϕ with $V(\phi) = 1$ is generally harmless, especially so when V is general. Errors in specifying V—which sentences V determines as true and which as false—are easy to spot. For example the verity inferences (logical consequence relations)

$$V(A \leftrightarrow BC) = 1, \ V(A) = 1 \vDash V(B) = 1$$
$$V(A \leftrightarrow (B \vee D)) = 1, \ V(\neg A) = 1 \vDash V(B) = 0$$

are both valid. However when taken together they, i.e., the four premises, are inconsistent—obviously one can't have a verity function that assigns 1 to both A and $\neg A$.

But with probability logic inconsistencies in assigning P values[14] may not be so obvious. We can illustrate this with the so-called "statistical paradox" whose puzzling feature occasioned considerable discussion in the philosophical literature of the 1950's and 1960's. (See *Suppes* 1966 and references therein.)

The paradox is stated by Suppes (*1966*, p. 49) as follows:

(1) The probability that Jones will live at least fifteen years given that he is now between fifty and sixty years of age is r. Jones is now betwen fifty and sixty years of age. Therefore the probability that Jones will live at least fifteen years is r.

(2) The probability that Jones will live at least fifteen years given that he is now between fifty-five and sixty-five years of age is s. Jones is now between fifty-five and sixty-five years of age. Therefore, the probability that Jones will live at least fifteen years is s.

Assuming what seems reasonable, that r should be larger than s, results in the paradox.

Letting $J_{\geq 15}$, L_{50-60}, L_{55-65} replace the sentences involved in these two statements, then the inferences (1) and (2) stated as probability logical consequences are:

$$P(J_{\geq 15} \dashv L_{50-60}) = r, \; L_{50-60} \vDash P(J_{\geq 15}) = r \qquad (3)$$

$$P(J_{\geq 15} \dashv L_{55-65}) = s, \; L_{55-65} \vDash P(J_{\geq 15}) = s. \qquad (4)$$

First of all, as we have defined logical consequence for probability logic, (3) and (4) are not properly constructed since each contains a syntactic expression where only semantic ones ought to be. This improper mixing of languages can be corrected by having P as a probability function determined by a probability model which also includes both L_{50-60} and L_{55-65} having P value 1. Then restating (3) and (4) we obtain

$$P(J_{\geq 15} \dashv L_{50-60}) = r, \; P(L_{50-60}) = 1 \vDash P(J_{\geq 15}) = r \qquad (5)$$

$$P(J_{\geq 15} \dashv L_{55-65}) = s, \; P(L_{55-65}) = 1 \vDash P(J_{\geq 15}) = s. \qquad (6)$$

---

[14]We are here referring to conditional probability logic but will be using the simple letter 'P' instead of the 'P^*' of §1.5

Taken separately (5) and (6) are valid consequence relations, being instances (for $a \in [0, 1]$) of

$$P(\psi \dashv \phi) = a, \ P(\phi) = 1 \vDash P(\psi) = a. \tag{7}$$

This is clear since no matter what in terms of atomic sentences the logical structure of $\psi \dashv \phi$ and ϕ are (ϕ being L_{50-60} in (5) and L_{55-65} in (6)), the logical consequence (7) holds in either case whatever the probability function P may be. But the implicit assumption is, or ought to be, that the P in (5) and in (6) refer to the same probability function and that all four premises involved in (5) and (6) are referring to this same probability function. But if this is the case then the paradox disappears as we have:

THEOREM 1.61. *If the premises of (5) and (6) hold for a common probability function P then $r = s$.*

PROOF. Follows, as we shall see, from the Lemmas 1, 2 and 3 stated below.   □

The argument for the Lemmas is quite simple but is being spelled out to emphasize its formal nature—as befits something being called a logic.

We have the (unstated) material assumptions that

$$L_{50-60} \equiv (L_{50-55} \vee L_{55-60})$$
$$L_{55-65} \equiv (L_{55-60} \vee L_{60-65})$$

with the three distinct disjuncts L_{50-55}, L_{55-60}, and L_{60-65} on the right being mutually exclusive. This gives:

LEMMA 1. $(L_{50-60} \wedge L_{55-65}) \equiv L_{55-60}$.

Next we have

LEMMA 2. *If* $P(L_{50-60}) = 1$ *and* $P(L_{55-65}) = 1$ *then*

(i) $P(L_{55-60}) = 1$, *and*

(ii) $P(L_{50-55}) = P(L_{60-65}) = 0$.

PROOF. For part (i), since

$$P(L_{55-60} \vee L_{55-65}) \geq P(L_{55-65})$$

and by hypothesis $P(L_{55-65}) = 1$, then

$$P(L_{55-60} \lor L_{55-65}) = 1. \tag{1}$$

Now from Lemma 1

$$P(L_{55-60}) = P(L_{50-60} \land L_{55-65})$$

so that, making use of (d) in §1.2, we have

$$= P(L_{50-60}) + P(L_{55-65}) - P(L_{50-60} \lor L_{55-65})$$

from which, by use of both parts of the hypothesis and (1),

$$= 1 + 1 - 1 = 1. \tag{2}$$

To establish part (ii) of Lemma 2 start with the first part of its hypothesis, obtaining

$$1 = P(L_{50-60})$$
$$= P(L_{50-55}) + P(L_{55-60})$$

so that, by (2),

$$= P(L_{50-55}) + 1.$$

Hence $P(L_{50-55}) = 0$. A similar argument shows that $P(L_{60-65}) = 0$. ☐

LEMMA 3. *If* $P(L_{50-60}) = 1$ *and* $P(L_{55-65}) = 1$, *then*

$$P(J_{\geq 15} \dashv L_{55-60}) = P(J_{\geq 15} \dashv L_{55-65}).$$

PROOF. For the two conditional probabilities involved in the conclusion, looking at just the numerators in the definition of a conditional probability (§1.5 (ii))above), both denominators here being equal, we have

$$P(J_{\geq 15} L_{50-60}) = P(J_{\geq 15}(L_{50-55} \lor L_{55-60}))$$
$$= P(J_{\geq 15} L_{50-55}) + P(J_{\geq 15} L_{55-60})$$
$$= 0 + P(J_{\geq 15} L_{55-60}),$$

since, by Lemma 2, $P(L_{50-55}) = 0$.

Similarly,

$$P(J_{\geq 15} L_{55-65}) = P(J_{\geq 15} L_{55-60}).$$

Hence as the Lemma contends the conditional probabilities in the statement
are equal.   □

The standard resolution of the paradox according to Suppes (*loc. cit.*)
is that:

> ... The inferences in question are illegitimate because the total
> evidence available has not been used in making the inference. Taking
> the two premises together we know more about Jones than either
> inference alleges, namely, that he is between fifty-five and sixty years
> of age.

In a sense our resolution, expressed by Theorem 1.61, is the same though
the justification we give for what is legitimate or illegitimate is via proof
based on explicitly formulated conditional-probability logic principles.

§1.7. Combining evidence

A topic of interest throughout the eighteenth and the first half of the
nineteenth century involved aspects of both probability and logic: How is
one to combine two items of evidence for a conclusion, each of which separately impart a probability for the conclusion, so as to have a probability
for the conclusion involving both items? The topic engendered considerable
controversy and confusion. (See SPL, §2.6, for some history.) One of the
reasons for this, among others, was the failure to distinguish between 'probability of a (verity) conditional' and 'conditional probability', i.e., between
$P(\phi \to \psi)$ and $P(\psi \dashv \phi)$.

Apparently Boole was the first to point this out and to formulate the
problem of combining evidence (using modern symbolism introduced after

Boole's time) as

$$\text{Given:} \quad P(z\,|\,x) = p, \quad P(z\,|\,y) = q$$
$$\text{Find:} \quad P(z\,|\,xy).$$

That is, given the conditional probability of z with respect to x and also that of z with respect to y, find the conditional probability of z with respect to the conjunction xy.[15]

We gain some insight into the problem by depicting the desideratum probability $P(z\,|\,xy) = P(xyz)/P(xy)$ in a probabilistic Venn diagram whose rectangle has unit area and with the probability of an event equal to the area of the region depicting the event. The diagrams of Figure 1 show that it is possible for $P(z\,|\,xy)$ to have the extreme values 0 and 1, as well as any value in between.

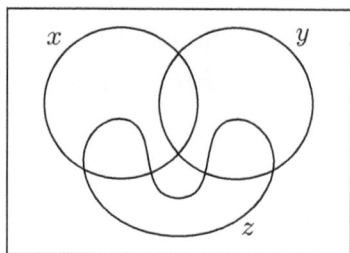

$$\frac{P(xyz)}{P(xy)} = 0$$

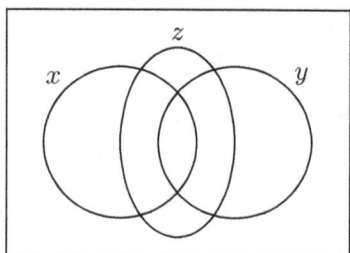

$$\frac{P(xyz)}{P(xy)} = 1.$$

FIGURE 1.

Could the additional premises $P(z\,|\,x) = p$ and $P(z\,|\,y) = q$ sufficiently

---

[15]Contained in the memoir *Boole* 1857, reprinted as Essay XVI in *Boole* 1952, of which the relevant pages are 355–367. On p. 356 of this book in the display labeled (1) there is a typographical error. The first numerator should be 'Prob xz'.

restrict the question so as to produce a useable result i.e., a value for $P(z|xy)$?

Before addressing ourselves to this question we describe Boole's "solution".

Boole reframes the problem in two respects: first of all by expressing the premise conditions linearly in terms of unconditional probabilities with the introduction of two new parameters, c and c', so that, e.g., in place of $P(z\,|\,x) = p$ he writes the two equations $P(x) = c$, $P(xz) = cp$, and a similar pair for the other premise; secondly, for the conditional probability $P(z\,|\,xy)$, which equals $P(xyz)/P(xy)$, he writes

$$\frac{P(xyz)}{P(xyz) + P(xy\bar{z})} \tag{1}$$

and determines the two unknown terms in this quotient by solving the linear problem

$$\text{Given:} \quad P(x) = c, \quad P(xz) = cp$$
$$P(y) = c', \quad P(yz) = c'q$$
$$\text{Required:} \quad P(xyz)$$

and then another with the same given but with Required: $P(xy\bar{z})$. The so obtained $P(xyz)$ and $P(xy\bar{z})$ are put into (1) and taken to be his value for $P(z\,|\,xy)$.

No doubt on account of its highly complicated appearance Boole checks his result with various values of p and q, one of these being $p = q = 1/2$, for which he obtains for $P(z\,|\,xy)$ the value $1/2$. That this is an incorrect reformulation can be shown by the following example.

Take $c = P(x) = .6$, $c' = P(y) = .6$ and assign to the eight constituents the following values:

$$P(xyz) = P(x\bar{y}\,\bar{z}) = P(\bar{x}yz) = .2$$
$$P(xy\bar{z}) = P(x\bar{y}z) = P(\bar{x}yz) = .1$$
$$P(\bar{x}\,\bar{y}z) = P(\bar{x}\,\bar{y}\,\bar{z}) = .05$$

It can then be seen (a probabilistic Venn diagram is helpful) that

$$p = P(xz)/P(x) = .5, \quad q = P(yz)/P(y) = .5$$

and that, differing from Boole's contention,

$$P(z \mid xy) = \frac{P(xyz)}{P(xy)} = \frac{.2}{.3} = \frac{2}{3}.$$

In our conditional-probability logic context Boole's problem here can be stated, in general form, as:

Given sentential formulas ψ, ϕ_1, ϕ_2 expressed in terms of atomic sentences what is the optimal β so that

$$P(\psi \mid \phi_1) = p, \ P(\psi \mid \phi_2) = q \vDash P(\psi \mid \phi_1 \phi_2) \in \beta?$$

Solving the problem as a conditional-probability logic consequence entails expressing $P(\psi \mid \phi_1 \phi_2)$ in terms of probabilities of constituents which, for the case at hand is a *quotient* of linear forms. Finding optimal bounds for a quotient requires the use of linear fractional programming techniques. These techniques were only introduced in the 1960's (*Charnes and Cooper* 1962) which, of course, were not available to Boole. Applying these techniques we found (SPL, Theorem 5.45) that no matter what non-zero values $P(A_3 \mid A_1)$ and $P(A_3 \mid A_2)$ may have, $P(A_3 \mid A_1 A_2)$ can still have any value in $[0, 1]$, i.e., $\beta = [0, 1]$ is, with these premises, the optimal value for $P(A_3 \mid A_1 A_2)$. Clearly, more information added to the premises is needed to narrow the interval. We conclude this section (and chapter) with an interesting example of this.

The topic of combining evidence was revived in a contenporary paper (*Saunders-Meyer-Wu* 1999) in connection with a sensational trial of one accused of a double murder. The method used in their paper to find the probability of an event on combined evidence was quite different. The following is a summary.

The authors use I for the event of innocence and I' for that of guilt of the defendent. While there were several items of evidence involved, they limit the discussion to two: let M_1 be that the blood drop found at the murder scene was consistent with its being that of the defendent, and M_2 that of the blood found on a sock of the defendent being consistent with its being from one of the two victims. There are objective ways of estimating $P(M_1 \mid I)$, $P(M_2 \mid I)$ and the prior probabilities $P(I)$, $P(I')$. Adjoining these data as premises they also adjoin as a premise a property expressing

"M_1 and M_2 are more strongly associated given I' than given I" which they represent by the inequation

$$\frac{P((M_2 \mid M_1) \mid I')}{P(M_2 \mid I')} \geq \frac{P((M_2 \mid M_1) \mid I)}{P(M_2 \mid I)}. \tag{2}$$

(Note the occurrence here, not normally seen in standard probability theory, of a conditional within a conditional though formally treatable in our conditional probability logic. See §1.5 above.) From (2) the authors show that

$$P(I \mid M_1 M_2) \leq \frac{P(I)}{P(I')} \cdot \frac{P(M_1 \mid I)}{P(M_1 \mid I')} \cdot \frac{P(M_2 \mid I)}{P(M_2 \mid I')}$$

from which, on substituting the data values, they obtain

$$P(I \mid M_1 M_2) \leq 8.65 \times 10^{-9}, \tag{3}$$

with the conclusion "guilty beyond reasonable doubt".

Obtaining a bound on $P(I \mid M_1 M_2)$ via a conditional-probability logical consequence relation using (2) as a premise is not possible; for when (2) is expressed in terms of constituents on I, M_1, M_2, the terms obtained are highly non-linear. As a replacement for (2) we introduced

$$P(I' \to (M_1 \leftrightarrow M_2)) \geq P(I \to (M_1 \leftrightarrow M_2)), \tag{4}$$

that is, the probability that guilt implies the equivalence of the two blood spot findings is greater than the probability that innocence does. It may be of interest to see a formulation of the linear fractional programming problem whose solution yields a value for the conditional probability of innocence given both items of evidence, with (4) instead of (2) as one of the premises.

Let a, c_1, c_2 be the reasonable estimates for $P(I)$, $P(M_1 \mid I)$, $P(M_2 \mid I)$ (e.g., $P(I)$ = chance that a randomly chosen resident of California is innocent) so that $P(IM_1) = c_1 a = b$ and $P(IM_2) = c_2 a = c$ are given numerical values for these probabilities. Taking $k_1, k_2, \ldots, k_8$ to be the probabilities of the eight components

$$M_1 M_2 I, \ M_1 M_2 I', \ M_1 M_2' I', \ M_1' M_2 I', \ M_1 M_2' I, \ M_1' M_2 I, \ M_1' M_2' I, \ M_1' M_2' I',$$

the premises are readily seen to be expressed by the equations

$$k_1 + k_3 + k_5 + k_7 = a$$
$$k_1 + k_3 = b$$
$$k_1 + k_5 = c \qquad (5)$$
$$\sum_{k=1}^{8} k_i = 1 \quad (k_i \geq 0)$$

plus an additional premise expressing (4) which, after a bit of logic on the expressions, becomes

$$k_2 + k_8 \geq k_1 + k_7.$$

Since the prior probability $P(I)$ is extremely close to 1 we have $k_8 = P(M_1'M_2'I') = 1 - P(I \vee M_1 \vee M_2) \approx 0$ and the preceding equation is replaceable by

$$k_2 \geq k_1 + k_7. \qquad (6)$$

Making use of this inequality we have for the probability of the conclusion

$$P(I \,|\, M_1 M_2) = \frac{k_1}{k_1 + k_2} \leq \frac{k_1}{2k_1 + k_7}. \qquad (7)$$

Then, with the given numerical values for a, b, c, from premises (5) plus (6) by the techniques of linear fractional programming (as illustrated in §5.4 of SPL), we[16] obtained

$$P(I \,|\, M_1 M_2) \leq 1.47 \times 10^{-10},$$

which is a slightly tighter upper bound—implying guilt a bit more strongly—than (3).

---

[16]This is a genuine plural, not an editorial "we". The problem, formulated as a linear fractional programming problem, was solved by Max Hailperin using the computer program *Mathematica*.

CHAPTER 2

LOGIC WITH QUANTIFIERS

§2.0. Ontologically neutral (ON) languages

This preliminary section is a brief sketch orienting the reader to a new way of viewing quantifier language, i.e., a language with quantifications. Exact formulation is given in the next section.[1]

The quantifier language (and logic based on it) to be introduced in this Chapter will serve as a basis for constructing our probability logic on quantifier language[2]. This probability logic will be much simpler than one based on a predicate language such as given, for example, in *Gaifman* 1964, §2.

To state this formal language we begin with standard sentential logic's formal language consisting of

(a) a finite set of atomic sentence symbols, here $A_0, A_1, \ldots, A_n$, where n is a numeral (we say "numeral" not "number" as we are referring to a subscript, i.e., a symbol on the letter A, and

(b) formulas which are appropriately defined finite combinations of these sentence symbols with symbols for logical constants (e.g., $\neg, \vee, \wedge$).

Although n is finite it can be taken as large as context requires, namely, equal to the number of atomic sentences being used in a logical investigation.

Ontologically neutral language is an extension of that on which sentential logic is based in two respects:

---

[1] Initially ON language was presented in *Hailperin* 1997. The present version is now enlarged and includes what the earlier version did not, namely what corresponds in predicate language to functions of individuals.

[2] Thinking of the general form of a language, we shall usually be using the singular form of the word 'language'.

1. The set of atomic sentence symbols is not only *enlarged in number to denumerably* many but also in complexity, becoming atomic sentence expressions. Here the subscript can be not only a numeral but also have a complex structure: it can also be an index (a variable), or one of a set of (symbols for) primitive recursive functions (specific and explicit for a given ON language); furthermore the argument places of these primitive recursive functions can be occupied by (i) a numeral, (ii) an index, or (iii) a primitive recursive function with arguments that are either numerals or indices. (All this is spelled out in our next section.) When all the arguments in such a subscript are numerals then, in virtue of the functions being primitive recursive, the subscript has an explicitly calculable numeral value.

2. Additionally to this extension of the notion of a sentential atomic sentence, ON language includes quantifier symbols that *semantically* function to generalize conjunction and disjunction of denumerably many formulas. We say "semantically" as ON language, the same as ordinary human language, doesn't have explicitly given sentences with more than a finite number of component symbols, and hence can't explicitly display all terms of a denumerable conjunction or disjunction of formulas.[3]

In comparing ON language with first-order predicate language with regard to semantics we shall see a pronounced difference in the concept of a model. In the case of predicate language a model consists of a non-empty set (called the "universe") and an interpretation over the universe for (i) the constants as specified elements of the universe, (ii) interpretations for the predicate symbols as predicates over (i.e., applying to elements of) the universe, and (iii) for the function symbols functions on elements of the universe with values that are in the universe. There are no conditions imposed on the universe except that it be non-empty—it can even be non-denumerable. Likewise interpretations for the functions are unrestricted.

In contrast, for an ON language there are only a countable (finite or denumerable) number of atomic sentences, and a model is simply an assignment of a truth value to each atomic sentence of the language; nothing

---

[3]There are some mathematical logicians who admit languages that do. For example *Scott and Krauss* 1966, a paper on probability logic, has (p. 221): "The expressions of $\mathcal{L}$ [an "infinitary" language] are defined as transfinite concatenations of symbols of length less than ω_1 [ω_1 is the first uncountable ordinal]. . . ."

is said about its linguistic structure or what the atomic sentence is about. Hence our epithet "ontologically neutral" describing the languages and their logic. It will be shown (Theorem 2.11) that a model specifies a unique truth value for each sentence of a given ON language.

§2.1. Syntax and semantics of ON logic

The general form of an ON language $\mathcal{Q}$ $[=\mathcal{Q}(p_1,\ldots,p_r;q_1,\ldots,q_s)]$ for which ON logic is formulated is as follows. Its symbols are

(a) numerals: $|, ||, |||, \ldots, \overbrace{||\cdots|}^{n+1}, \ldots$; abbreviated to $0, 1, \ldots, n, \ldots$ (borrowing from mathematics the symbols used for numbers),

(b) indices: $i_1, i_2, \ldots, i_n, \ldots$ (the subscripts on the i's here and on the p's and q's in (c) are to be taken simply as distinguishing tags, not the formal numerals of (a)), and the order displayed here is referred to as the "alphabetical",

(c) two finite sets of expressions $p_1, \ldots, p_r$ and $q_1, \ldots, q_s$ (varying with the language) denoting in virtue of their construction, primitive recursive (p.r.) functions from numerals to numerals, each with a specified number of argument places (understood, though not indicated in the notation). The first set, $\mathcal{Q}$'s *primary* (p.r.) functions, are to

    (c_1) be defined for all numerals as arguments

    (c_2) have mutually exclusive ranges, the union of these ranges being a set of numeral expressions representative of the entire set of numerals (see below for amplification), and

    (c_3) be one-to-one on the numerals. The second set, $q_1, \ldots, q_s$, $\mathcal{Q}$'s *subordinate* (p.r.) functions are unrestricted except that they are to occur only as arguments in primary functions. Any such occurrence will be referred to as a *subscript-term*.

(d) the letter A with a subscript that is one of the $p_1, \ldots, p_r$ with its argument place(s) occupied by either numerals or indices or a subscript-term, the latter a subordinate function whose argument place(s) are occupied by indices or numerals,

(e) sentential connectives: $\neg, \wedge, \vee, (\rightarrow, \leftrightarrow$ defined as usual)

(f) *and*-quantifiers: $(\bigwedge i), (\bigwedge j), \ldots,$

or-quantifiers: $(\bigvee i), (\bigvee j), \ldots$

$(i, j, \ldots$ being indices),

(g) left and right parentheses: (,).

Briefly, primitive recursive (p.r.) functions are those obtainable by use of a general form of definition by induction.[4] Being notationally specified they can be meaningfully used as part of an inscription (i.e., in a subscript on A) when their argument places are filled either by numerals or by indices. The letter A with a subscript (as defined in (d)) is an *atomic formula*, and when in the subscript there are no indices, i.e., when the subscript is a *numeral expression*, it is an *atomic sentence*. It follows from (c_1)–(c_2) that for a *given* ON language each numeral expression is uniquely associated with a numeral, its value, and, conversely, given a numeral the particular p.r. function p_i and arguments in it that produce this numeral are effectively retrievable. (See §2.3 below for a specific EXAMPLE of an ON language where this can be readily seen.) In other words, an ON language's $p_1, \ldots, p_r$ specify r distinct non-overlapping sets of numerals whose union is the entire set of all numerals, thus determining r independent families of the entire set of atomic sentences. The subordinate functions serve to extablish complex interrelations, similar to the use of functions of individuals in a predicate language.

The definition of a *formula* for ON language is similar to that for first-order predicate language except for having subscripted A's in place of predicate expressions and with indices in place of individual variables in quantifiers. We take over the usual abbreviations and conventions used with predicate logic such as *free, bound, scope,* and safeguards against confusion of bound indices. Thus a closed ON formula is one without free indices.

Since any finite set of atomic sentences of a $\mathcal{Q}$, in combination with sentential connectives behaves like a formula of an $\mathcal{S}$, ($\mathcal{S}$ as in Chapter 1) it will be convenient to describe this by saying "$\mathcal{S}$ is contained in $\mathcal{Q}$".

A *verity model M* for an ON language $\mathcal{Q}$ is an assignment of a truth value, *true* or *false*, to each of $\mathcal{Q}$'s atomic sentences. Such an assignment determines a truth value for each of a $\mathcal{Q}$'s closed formulas. This is shown by

---

[4]See, e.g., *Kleene* 1952, §43 or *Smorynski* 1977, p. 831.

means of the notion of a verity function. As in sentential logic we employ
'1' and '0' to stand for *true* and *false,* and use the arithmetic properties of
the numbers 1 and 0 to express semantic properties. Let $\mathcal{Q}^{cl}$ denote the set
of closed formulas of $\mathcal{Q}$. A function V from $\mathcal{Q}^{cl}$ to the set of truth values
$\{0, 1\}$ is a *verity function* if, for any formulas ϕ, ψ, $(\bigwedge i)\chi$ and $(\bigvee i)\chi$ in $\mathcal{Q}^{cl}$,
it has the properties

(v$_1$) $V(\neg\phi) = 1 - V(\phi)$,

(v$_2$) $V(\phi \wedge \psi) = \min\{V(\phi), V(\psi)\}$,

(v$_3$) $V(\phi \vee \psi) = \max\{V(\phi), V(\psi)\}$,

(v$_4$) $V((\bigwedge i)\chi) = \min\{V(\chi(0)), \ldots, V(\chi(\nu)), \ldots\} = \min_{\nu}\{V(\chi(\nu))\}$,

(v$_5$) $V((\bigvee i)\chi) = \max\{V(\chi(0)), \ldots, V(\chi(\nu)), \ldots\} = \max_{\nu}\{V(\chi(\nu))\}$

where, for any numeral ν, $\chi(\nu)$ is the result of replacing each free occurrence
of the index i in χ by the numeral ν.[5]

We point out that in (v$_4$) and (v$_5$) of the definition of a verity function
we are enlarging our semantic "arithmetic" of the truth values, it being
assumed there that the min and max of an infinite sequence of truth values
is well-defined. This is readily justified since for an infinite sequence of 0's
and 1's there are three cases possible:

(s$_1$) the sequence consists of all 1's

(s$_2$) the sequence consists of all 0's

(s$_3$) there are 0's and 1's in the sequence.

In case (s$_1$) the max is 1, and and also the min; for case (s$_2$) the min is
0 and and also the max; and in case (s$_3$) the max is 1 and the min is 0.
Thus in any case there will be a defined max or min or both.

We now show that a model M, defined as assigning a truth value to the
atomic sentences of an ON language $\mathcal{Q}$, can be extended so as to be a verity
function V_M on $\mathcal{Q}^{cl}$, M then determining (via its V_M) a unique truth value
for any occurrence of each such formula and thereby providing semantic
meaning for the logical connectives and quantifiers of a closed formula, and
hence the formula's truth or falsity in the model.

---

[5]Introduction of the convenient abbreviation '$\chi(\nu)$' for '$\chi[\nu/i]$' is not to be taken as
a backdoor attempt to bring in numerals as arguments (of predicates). Context will
indicate which index the ν is replacing —as, for example, in (v$_4$) and (v$_5$) it is to be
understood that i is being replaced.

THEOREM 2.11. *To each verity model M of an ontologically neutral language $\mathcal{Q}$ there is a unique verity function V_M extending M to all of $\mathcal{Q}^{cl}$.*

PROOF. Let M be a model for $\mathcal{Q}$, i.e., a function assigning a value, 0 or 1, to each of $\mathcal{Q}$'s atomic sentences. We define V_M by strong induction on the number of occurrences of logical operators (connectives or quantifiers) in a closed formula ϕ.

BASIS. Let ϕ have no logical operators.

Then ϕ is an atomic sentence A_ϵ, ϵ a numeral expression. We set

$$V_M(A_\epsilon) = M(A_\epsilon).$$

INDUCTION STEP. For closed formulas of the form $\neg\phi$, $\phi \wedge \psi$, $\phi \vee \psi$ the equations

$$V_M(\neg\phi) = 1 - V_M(\phi)$$
$$V_M(\phi \wedge \psi) = \min\{V_M(\phi), V_M(\psi)\}$$
$$V_M(\phi \vee \psi) = \max\{V_M(\phi), V_M(\psi)\}$$

define their V_M value in terms of those of closed formulas with fewer operators. For closed formulas of the form $(\bigwedge i)\phi$ and $(\bigvee i)\phi$ we define their V_M value by setting

$$V_M((\bigwedge i)\phi) = \min_\nu\{V_M(\phi(\nu))\}$$
$$V_M((\bigvee i)\phi) = \max_\nu\{V_M(\phi(\nu))\}.$$

As $(\bigwedge i)\phi$ and $(\bigvee i)\phi$ are closed, so are each of the instances $\phi(\nu)$. And being closed and with one less operator the $\phi(\nu)$ have, by hypothesis of induction, a defined V_M value. This completes the inductive definition of V_M. A simple proof by induction establishes that any two such functions V_M and V'_M both extending M to all of $\mathcal{Q}^{cl}$ must be identical. $\square$

We say a closed formula ϕ is *true in a model M* (for a given $\mathcal{Q}$) if, for its associated V_M, $V_M(\phi) = 1$. It is *valid* if true in every model M of $\mathcal{Q}$.

Two (closed) formulas ϕ and ψ of an ON language are *semantically equivalent*, symbolized by '$\phi \equiv \psi$', if the equivalence $\phi \leftrightarrow \psi$ is valid or, equivalently, if $V(\phi) = V(\psi)$ in every verity model V.

The following examples should be of help in acquiring a feel for this new quantifier language.[6]

EXAMPLE 1. Let $\mathcal{Q}(p(i); q(i))$ be an ON language which has one primary function $p(i) = i$ and a subordinate function $q(i) = i + 2$. Then, for ϕ a formula with only one free index i,

$$\phi(p(0))\phi(p(1))(\bigwedge i)\phi(p(q(i))) \equiv (\bigwedge i)\phi(p(i))$$

or, since $p(0) = 0$, $p(1) = 1$, $p(q(i)) = i + 2$, written more simply in abbreviated form,

$$\phi(0)\phi(1)(\bigwedge i)\phi(i + 2) \equiv (\bigwedge i)\phi.$$

To see that this holds we have by use of (v_2) and (v_4) above

$$V(\phi(0)\phi(1)(\bigwedge i)\phi(i + 2)),$$
$$= \min\{V(\phi(0)), V(\phi(1)), \min\{V(\phi(2)), \ldots, \}\}$$
$$= \min\{V(\phi(0)), \ldots, V(\phi(\nu)), \ldots\}$$
$$= V((\bigwedge i)\phi).$$

EXAMPLE 2. Let $\mathcal{Q}(p(i); q(i,j))$ be an ON language with primary function $p(i) = i$ and a subordinate one $q(i,j) = i + j$.

Here the condition (c_2)—that the primary functions of a language have mutually exclusive ranges—is satisfied since the range of p, the only primary function of $\mathcal{Q}$, is the set of all numerals. As for the restriction (c_3), that the primary functions be one-to-one on the numerals, that too is also satisfied; although, e.g., $q(2,3)$ and $q(1,4)$ have, for the q of this example, distinct argument pairs but equal values, it is only the primary functions that are restricted by (c_3), and in this case $A_{p(q(2,3))} = A_{p(5)} = A_5$, and $A_{p(q(1,4))} = A_{p(5)} = A_5$, so that the primary function here, p, does meet with the restriction.

This example illustrates that a formula $A_{p(i,j)}$ can either have meaning or be meaningless depending on the ON language one is using: For $\mathcal{Q}_1 =$

---

[6]To simplify appearances we are using juxtaposition for '∧' and raising subscripts up to "line level" as if they were "arguments".

$\mathcal{Q}_1(p(i,j);)$ with $p(i,j) = i+j$ it is meaningless, $\mathcal{Q}_1$ not being in accordance with specifications of an ON language. But for $\mathcal{Q}_2 = \mathcal{Q}_2(p(i); \ q(i,j))$, p being the identity $p(i) = i$ and $q(i,j) = i+j$ it is acceptable, as our above cited Example 2 shows.[7]

A set of closed ON formulas is *satisfiable* if there is a model M such that for its associated V_M every element of the set has V_M value 1. The set is *unsatisfiable* if there is no such M, i.e., if for every M its associated V_M has value 0 for at least one element of the set.

EXAMPLE 3. Let ϕ be a formula with one free index i. Then the infinite set

$$\Gamma = \{(\textstyle\bigwedge i)\phi, \neg\phi(0), \ldots, \neg\phi(\nu), \ldots\}$$

is unsatisfiable.

PROOF. For any M, by (v_4) of the definition of a verity function,

$$V_M((\textstyle\bigwedge i)\phi) = \min\{V_M(\phi(0)), \ldots, V_M(\phi(\nu)), \ldots\}. \tag{1}$$

Case (i). $V_M((\bigwedge i)\phi) = 1$.

Then by (1), for each ν, $V_M(\phi(\nu)) = 1$ and so for each ν $V_M(\neg\phi(\nu)) = 0$. Hence Γ is unsatisfiable.

Case (ii). $V_M((\bigwedge i)\phi) = 0$.

Then by (1) there is some ν_0, such that $V_M(\neg\phi(\nu_0)) = 0$. Hence again Γ is unsatisfiable. Since this exhausts the possibilities, Γ is unsatisfiable.□

Just as with sentential probability logic, obtained as a generalization of sentential verity logic, we shall be presenting quantifier probability logic as a generalization of quantifier verity logic. The generalization can be made more natural looking if we rephrase (v_4) and (v_5) of the definition of a verity function for quantifier formulas. To this end we use "finite quantifiers". As an abbreviation for the $(\nu + 1)$-fold conjunction '$\phi(0) \wedge \cdots \wedge \phi(\nu)$' we write '$\bigwedge_{i=0}^{\nu}\phi$', and similarly '$\bigvee_{i=0}^{\nu}\phi$' for the $(\nu + 1)$-fold disjunction. Then in place of (v_4) one can have

(v_4') $V((\bigwedge i)\chi) = \lim_{\nu\to\infty} V(\bigwedge_{i=0}^{\nu}\chi)$

in virtue of

---

[7]Predicate language isn't faced with this need for a symbolic distinction as $P(i,j)$ and $P(i+j)$ are obviously distinct, the first being a two-argument predicate (relation) while the second is a one-argument predicate whose argument is a two-argument function.

THEOREM 2.12.

$$\lim_{\nu \to \infty} V(\textstyle\bigwedge_{i=o}^{\nu}\chi) = \min\{V(\chi(0)), \ldots, V(\chi(\nu)), \ldots \}.$$

PROOF. If $V(\chi(\nu)) = 0$ for some $\nu = \nu_0$ then $V(\bigwedge_{i=0}^{\mu}\chi) = 0$ for all $\mu \geq \nu_0$, and so both $\lim_{\nu \to \infty} V(\bigwedge_{i=0}^{\nu}\chi)$ and $\min\{V(\chi(0)), \ldots, V(\chi(\nu)), \ldots \}$ are 0. While if $V(\chi(\nu))$ is not 0 for some ν_0 it is 1 for all ν and again we have equality.   □

Similarly, another way of writing (v_5) of the definition is

$$V((\textstyle\bigvee i)\phi) = \lim_{\nu \to \infty} V(\textstyle\bigvee_{i=0}^{\nu}\phi).$$

As another example, taking χ in (v_4) to be $(\bigwedge j)\phi(i, j)$,

$$
\begin{aligned}
V((\textstyle\bigwedge i)(\textstyle\bigwedge j)\phi(i, j)) &= \min_{\nu}\{V((\textstyle\bigwedge j)\phi(\nu, j))\} \\
&= \min_{\nu}\{\min_{\mu}\{V(\phi(\nu, \mu))\}\} \\
&= \lim_{\nu \to \infty} V(\textstyle\bigwedge_{i=0}^{\nu}\textstyle\bigwedge_{j=0}^{\nu}\phi(i, j)),
\end{aligned}
$$

since each expression on the right of the equals sign has value 0 if and only if there is a pair of numerals (ν_0, μ_0) such that $V(\phi(\nu_0, \mu_0)) = 0$. Similarly, for more than two $\bigwedge$-quantifiers,

$$V((\textstyle\bigwedge i_1) \cdots (\textstyle\bigwedge i_n)\phi(i_1, \ldots, i_n)) = \lim_{\nu \to \infty} V(\textstyle\bigwedge_{i_1=0}^{\nu} \cdots \textstyle\bigwedge_{i_n=0}^{\nu}\phi(i_1, \ldots, i_n)).$$

By replacing in this equation ϕ with $\neg\phi$ and using simple logical and mathematical transformations one has

$$V((\textstyle\bigvee i_1) \cdots (\textstyle\bigvee i_n)\phi(i_1, \ldots, i_n)) = \lim_{\nu \to \infty} V(\textstyle\bigvee_{i_1=0}^{\nu} \cdots \textstyle\bigvee_{i_n=0}^{\nu}\phi(i_1, \ldots, i_n)).$$

§2.2. Axiomatic formalization of ON logic

Sentential probability logic has a fundamental property that $P(\phi)$ is 1 if ϕ is truth-functionally valid (§1.2, P1). But to use ON's validity for the corresponding property in defining probability on ON quantifier language (to be introduced in the next chapter) is not as convenient as provability in a formal system, and this is what we shall be using. Almost any first-order predicate axiomatization can be pressed into the service if appropriately revamped: using schematic letters for formulas, replacing $\forall$ by $\bigwedge$, $\exists$ by $\bigvee$, variables by indices and constants by numerals. This last replacement doesn't force numerals to play the role of constants in ON language; e.g., supposing $A_{p(2)}$ is a formula in ON language, it need not be saying anything about '2'. The numeral which $p(2)$ equals serves to pick out one of the atomic sentences $A_0, A_1, \ldots, A_\nu \ldots$, namely the one whose subscript value is equal to $p(2)$, which sentence need not have any connection with 2, i.e., need not be saying anything about 2 except if a specific model happens to be calling for it.

As an axiomatization for ON logic we choose the following which is based on a predicate logic axiomatization due to Herbrand, slightly modified.[8]

Axioms For ON Logic
The truth-functionally valid quantifier-free formulas.

Rules of Inference

R1. ($\bigwedge$-*quantifier introduction*) From a formula ϕ not containing an occurrence of a bound index i, infer $(\bigwedge i)\phi$.

R2. ($\bigvee$-*quantifier introduction*) From a formula ϕ not containing an occurence of a bound index j, infer $(\bigvee j)\phi'$, where ϕ' is like ϕ except for one or more occurrences of free i, or subscript term with free indices, replaced by j.

R3. (*Passage*) Rules for moving '¬' and quantifiers over formulas including 'relettering' of bound indices.[9]

---

[8]See *Herbrand* 1971, p. 175 and p. 192. In effect Herbrand showed that this axiomatization (i.e., its predicate logic form), which doesn't include *modus ponens*, is nevertheless equivalent to a "standard" one which does.

[9]It is assumed that these are familiar to the reader and need not be spelled out. Likewise omitted is a statement of the conditions to prevent confusion of bound indices,

R4. (*Simplification*) Replacement of a formula part of the form $\phi \vee \phi'$, ϕ' an alphabetic (indices) variant of ϕ, by ϕ.

To state that ϕ is a theorem of ON logic we shall write '⊢ ϕ'.

It is of particular interest to note that, *modus ponens* not being required, all proofs have the form of a linear succession of formulas of which the first is an axiom, and for the remaining, each one comes from its predecessor by a use of a rule of inference. This is quite handy in discussions about arrays of formulas constituting a proof.

Since there is no essential syntactic difference between first-order predicate logic and ON logic when atomic sentences are not explicitly involved, just about all standard material about first-order predicate logic can be taken over:

THEOREM 2.20. *When no explicit reference is made to atomic sentence structure all provable results about first-order logic can be taken over as ON logic results. For example: replacement of a formula by a logically equivalent one, change of bound variable (i.e., index), use of the Deduction Theorem, and the like.*

Later we shall have need for results about finite quantifiers. For convenience of reference we make a formal statement:

THEOREM 2.21. *All general properties of quantifiers, e.g., such as duality of $\bigwedge$ and $\bigvee$, interchange of quantifiers of like kind, rules of passage of quantifiers over formulas, and the like, carry over to finite quantifiers.*

For example this Theorem 2.21 justifies the following as theorems of ON logic:

$$\vdash \neg \bigvee_{i=0}^{n} \phi \leftrightarrow \bigwedge_{i=0}^{n} \neg \phi$$

$$\vdash \bigwedge_{i=0}^{n} \bigwedge_{j=0}^{m} \phi \leftrightarrow \bigwedge_{j=0}^{m} \bigwedge_{i=0}^{n} \phi$$

$$\vdash \bigwedge_{i=0}^{n} (\theta \vee \phi) \leftrightarrow \theta \vee \bigwedge_{i=0}^{n} \phi, \quad \text{if } \theta \text{ has no free } i.$$

---

similar to that of first-order predicate logic for bound individual variables.

§2.3. Adequacy of ON logic

Despite the substantial semantic differences between (formal) first-order predicate language and that of ON language, e.g., one allowing (ostensibly) for non-denumerable models the other not, on a symbol for symbol basis a correlation can be established in such a way that (statable) provable formulas correspond. As a simple illustration of this correlation, $A_{p(i,j)}$ correlates with $P(a_i, a_j)$, p being a primary p.r. function and P a symbol for a predicate (with its number of arguments specified), i correlating with a_i, j with a_j; likewise $A_{p(i,q(j))}$ with $P(a_i, f(a_j))$, q a subordinate p.r. function and f one on individuals. To justify this assertion requires entering into the details of provability for the respective formal languages.

Let $\mathcal{L}(P_1, \ldots, P_m; f_1, \ldots, f_r; a_1, a_2, \ldots, a_n, \ldots)$ be a (formal) first-order predicate language with symbols for predicates (the P's), function symbols (the f's) and constant symbols (the a's). Each of the predicate and function symbols are to have a specified number of argument places (not shown). With this $\mathcal{L}$ we associate an ON language $\mathcal{Q}(p_1, \ldots, p_m; q_1, \ldots, q_r)$ whose p.r. functions are as follows. Each p_k $(k = 1, \ldots, m)$, $\mathcal{Q}$'s primary functions, will have the same number of argument places as the predicate P_k and the values of each of these functions will range (respectively) over the m different residue classes of natural numbers modulo m (thus keeping the ranges of the p_k from overlapping). This is achieved by setting:

$$p_k(x_1, \ldots, x_{m_k}) = mN^{(m_k)}(x_1, \ldots, x_{m_k}) + k - 1,$$

where $N^{(n)}(x_1, \ldots, x_n)$ is a primitive recursive function mapping n-tuples of natural numbers (= numerals, for us) to the natural numbers (numerals).

As an EXAMPLE of this mapping take the case of three predicate symbols P_1, P_2, P_3 having, respectively, 1, 1 and 2 argument places, we would have the three p.r. functions:

$$p_1(x) = 3N^{(1)}(x) + 0 = 3x$$
$$p_2(x) = 3N^{(1)}(x) + 1 = 3x + 1$$
$$p_3(x, y) = 3N^{(2)}(x, y) + 2 = 3\left[\binom{x + y + 1}{2} + \binom{x}{1}\right] + 2.$$

In the specification of $p_3(x, y)$ the two-level components in parentheses are binomial coeficients, e.g.,

$$\binom{a}{b} = \frac{a!}{b!(a-b)!},$$

and one readily computes that, as x and y range over all the non-negative integers, the values of $p_1(x)$ are all the multiples of 3, those of $p_2(x)$ all the multiplies of 3 plus 1, and those of $p_3(x, y)$ all the multiples of 3 plus 2.

Returning to the correlation assertion, the p.r. functions $q_1, \ldots, q_r$, i.e., $\mathcal{Q}$'s subordinate functions, are to have the same number of argument places, respectively, as the functions $f_1, \ldots, f_r$, and be defined for all numerals as arguments, but otherwise unrestricted. We have:

THEOREM 2.31. *Let* $\mathcal{L}(P_1, \ldots, P_m; f_1, \ldots, f_r; a_1, \ldots, a_n, \ldots)$ *be a first-order language with predicate, function, and individual constant symbols as shown. Furthermore, let* $\mathcal{Q}(p_1, \ldots, p_m; q_1, \ldots, q_r)$ *be the ontologically neutral language associated with this* $\mathcal{L}$*, as described above. With each formula* ϕ *of* $\mathcal{L}$ *associate in one-to-one fashion a formula of* ϕ^* *of* $\mathcal{Q}$ *via the following associating rules:*

(i) *$\forall$ with $\bigwedge$, $\exists$ wth $\bigvee$;*

(ii) *distinct individual variables with distinct indices, e.g., x_n with i_n $(n = 1, 2, \ldots)$; predicate symbols $P_1, \ldots, P_m$ with primary p.r. functions $p_1, \ldots, p_m$; function symbols $f_1, \ldots, f_r$ with subordinate p.r. functions $q_1, \ldots, q_r$,*

(iii) *each predicate expression $P_k(v_1, \ldots, v_{m_k})$ by the letter A with subscript $p_k(v_1^*, \ldots, v_{m_k}^*)$, where v_l^* $(1 \leq l \leq m_k)$ is the index associated (by (ii)) with v_l if v_l is a variable, or is the numeral n if v_l is the individual constant a_n $(n = 1, 2, \ldots)$ or is a subscript term $q_l(w_1^*, \ldots, w_t^*)$ if v_l is the term $f_l(w_1, \ldots, w_t)$ and the w^* are the individuals or numerals corresponding to the individual variables or constants w in $f_l(w_1, \ldots, w_t)$.*

Then the list $\phi_1, \ldots, \phi_t$ (ϕ_t being ϕ) is a proof of ϕ in $\mathcal{L}$ if and only if $\phi_1^*, \ldots, \phi_t^*$ (ϕ_t^* being ϕ^*) is a proof of ϕ^* in $\mathcal{Q}$.

PROOF. The result is readily seen to hold by virtue of the identity (in schematic letters) between the formal logical system for $\mathcal{L}$ and that for $\mathcal{Q}$. Thus, for example a tautology (truth-functionally valid formula) remains a

tautology since the described replacements do not alter its truth-functional
form and the rules of inference as given in the preceding section are indif-
ferent to the replacements as described in (i)–(iii).   □

Needless to say, model theory for predicate logic and that for ON logic
are vastly different.

§2.4. Quantifier language with the suppositional

In this section we describe (in semantic terms) the language obtained
by adjoining to the (verity) ON language of §2.1 the suppositional connective. Later, in Chapter 4, it will serve in the development of conditional
probability for quantifier language.

Let $\mathcal{Q}^u$ denote the set of formulas obtained by augmenting $\mathcal{Q}$ with those
obtained by using the connective '⊣'. Taking the place of the suppositional
function U of sentential language is that of a $\mathcal{Q}$-*suppositional function* for
which we will continue to use the symbol U and drop the prefix '$\mathcal{Q}$-' from
'$\mathcal{Q}$-suppositional'. For $\mathcal{Q}^u$ the properties of the sentential connectives will
be the same as as listed in §1.4, namely

$$U(\neg\Phi) = 1 - U(\Phi)$$
$$U(\Phi \wedge \Psi) = \min\{U(\Phi), U(\Psi)\}$$
$$U(\Phi \vee \Psi) = \max\{U(\Phi), U(\Psi)\}, \tag{1}$$

though now Φ and Ψ stand for arbitrary closed formulas of $\mathcal{Q}^u$. To these
we add one for the suppositional

$$U(\Psi \dashv \Phi) = U(\Phi\Psi \vee \neg\Phi\mathbf{u})$$
$$= \max\{\min\{U(\Phi), U(\Psi)\}, \min\{1 - U(\Phi), u\}\}, \tag{2}$$

$\mathbf{u}$ being a formula that has the constant value u.[10]

---

[10]The kinship (and difference!) of '⊣' and '→' can be seen on comparing the first line
of (2) with the valid verity relation: $V(\phi \rightarrow \psi) = V(\phi\psi \vee \neg\phi\mathbf{1})$, with $\mathbf{1}$ being a constant
having the verity value 1.

Further we add semantic specifications for the quantifiers. Formally they are the same as for verity functions but with 'V' replaced with 'U'. Thus, Φ being a $\mathcal{Q}^u$ formula with one free index i,

$$U((\bigwedge i)\Phi) = \min\{U(\Phi(0)), \dots, U(\Phi(\nu)), \dots\}$$
$$U((\bigvee i)\Phi) = \max\{U(\Phi(0)), \dots, U(\Phi(\nu)), \dots\}. \tag{3}$$

That any infinite (as well as any finite) sequence which contains u's as well as 0's and 1's, such as depicted on the right in (3), does have a max and a min follows from the "arithmetical" properties ascribed to u (in §1.4) which, we are assuming, carry over.

A *model M for $\mathcal{Q}^u$* is, as with verity logic, an assignment of a verity value, 0 or 1, to its atomic sentences and now, additionaly, the value u to $\mathbf{u}$, which then serves to define a unique value, 0, u, or 1, to each closed formula of $\mathcal{Q}^u$ in virtue of the following:

THEOREM 2.41. *To each model M of a language $\mathcal{Q}^u$ there is a unique ($\mathcal{Q}$-)suppositional function U_M extending M (i.e., supplying a value, either truth value or u) to each closed formula of $\mathcal{Q}^u$.*

PROOF. Assuming the hypothesis, we establish the existence of a U_M by strong induction on the number of occurrences of logical operators (constants, connectives and quantifiers) in a formula.

BASIS. Let Φ have no logical operators.

(i) Φ is the constant $\mathbf{u}$. Then the set $U_M(\Phi) = u$.

(ii) Φ is an atomic sentence A_ϵ, ϵ a numeral expression. Then we set

$$U_M(A_\epsilon) = M(A_\epsilon).$$

INDUCTION STEP. For closed formulas of the form $\neg\Phi$, $\Phi \wedge \Psi$, $\Phi \vee \Psi$ equations (1) defined their U_M value in terms of those with fewer operators. As for the suppositional connective, from (2) we see that the U_M of $\Psi \dashv \Phi$ is expressible in terms of the U_M of Ψ and Φ, each of which has fewer operators than $\Psi \dashv \Phi$. As for quantifications, equations (3) show that, since the U_M of each of $\Phi(0), \dots, \Phi(\nu), \dots$ is defined—these formulas having one less operator than the quantification—then so also is the quantification. Finally a separate proof by induction, starting with the BASIS that any two U_M's being equal on the atomic sentences, have to be equal at each higher (construction) level, establishes their identity over every closed formulas. $\square$

Since formulas of $\mathcal{Q}^u$ not containing an occurrence of '$\dashv$' or **u** are the same as formulas of $\mathcal{Q}$, and have the same properties, we shall say "$\mathcal{Q}$ is a subset of $\mathcal{Q}^u$". Likewise, formulas of $\mathcal{Q}^u$ not containing quantifiers function as if they were in $\mathcal{S}^u$ where quantifiers are not involved, we shall say "$\mathcal{S}^u$ is a subset of $\mathcal{Q}^u$".

We illustrate the nature of this quantifier suppositional language with some simple examples. Here too, as with sentential suppositional language, we shall write '$\Phi \equiv_u \Psi$' to say that the formulas Φ and Ψ are *u-equivalent*, i.e., have the same values in any model. A straightforward proof by induction on the structure of a formula shows that replacement of a formula (part) by a *u*-equivalent one preserves *u*-equivalence.

For use in the following EXAMPLE 1 note that if in (2) above $U(\Phi) = u$, then

$$U(\Psi \dashv \Phi) = \max\{\min\{u, U(\Psi)\}, \min\{1 - u, u\}\}$$

and which, since (see §1.4) both $1 - u$ and $\min\{1 - u, u\}$ equal u,

$$= \max\{\min\{u, U(\Psi)\}, u\}$$
$$= u;$$

i.e. if its antecedent has the value u, then so does the suppositional.

EXAMPLE 1. Let Φ and $(\bigwedge i)\Psi$ be closed formulas of $\mathcal{Q}^u$. Then

(a) $(\bigwedge i)\Psi \dashv \Phi \equiv_u (\bigwedge i)(\Psi \dashv \Phi)$

(b) $(\bigvee i)\Psi \dashv \Phi \equiv_u (\bigvee i)(\Psi \dashv \Phi)$.

To see this note, for (a), that if $U(\Phi) = 1$, U an arbitrary suppositional function then

$$U((\bigwedge i)\Psi \dashv \Phi) = U((\bigwedge i)\Psi)$$
$$= \min\{U(\Psi(0)), \ldots, U(\Psi(\nu)), \ldots\}$$
$$= \min\{U(\Psi(0) \dashv \Phi), \ldots, U(\Psi(\nu) \dashv \Phi), \ldots\}$$
$$= U((\bigwedge i)(\Psi \dashv \Phi));$$

and if $U(\Phi) = 0$ then, the first line by (2)

$$U((\bigwedge i)\Psi \dashv \Phi) = u$$
$$= \min\{u, \ldots, u, \ldots\}$$
$$= \min\{U(\Psi(0) \dashv \Phi), \ldots, U(\Psi(\nu) \dashv \Phi), \ldots\}$$
$$= U((\bigwedge i)(\Psi \dashv \Phi)),$$

while if $U(\Phi) = u$ then by the note preceding the statement of this EXAM-PLE,

$$U((\bigwedge i)\Psi \dashv \Phi) = u;$$

and, further, for the right-hand side of (a),

$$U((\bigwedge i)(\Psi \dashv \Phi)) = \min\{U(\Psi(0) \dashv \Phi), \ldots, U(\Psi(\nu) \dashv \Phi), \ldots\}$$
$$= \min\{u, \ldots, u, \ldots\}$$
$$= u,$$

which establishes (a), the first of these two u-equivalences. The second, i.e. (b), follows similarly, with max in place of min. Note the resemblance of these u-equivalences to the corresponding ones for the verity conditional (no free i in Φ):

$$(\Phi \rightarrow (\bigwedge i)\Psi) \equiv (\bigwedge i)(\Phi \rightarrow \Psi)$$
$$(\Phi \rightarrow (\bigvee i)\Psi) \equiv (\bigvee i)(\Phi \rightarrow \Psi).$$

While there are similar equivalences in verity logic allowing for the movement of a quantifier out of the antecedent of an '$\rightarrow$' there aren't any for '$\dashv$'. Counterexamples of this are readily constructed. For example, let Ψ, $(\bigwedge i)\Phi$, $(\bigvee i)\Phi$ be such that for some U

$$U(\Psi) = 1, \quad U((\bigwedge i)\Phi) = 0, \quad U((\bigvee i)\Phi) = 1.$$

Then

$$U(\Psi \dashv (\bigwedge i)\Phi) = U(1 \dashv 0) = u,$$

and

$$U((\bigvee i)(\Psi \dashv \Phi)) = U((\bigvee i)(1 \dashv \Phi)) = 1.$$

Hence $\Psi \dashv (\bigwedge i)\Phi$ and $(\bigvee i)(\Psi \dashv \Phi)$ are not u-equivalent.

EXAMPLE 2. Let Ψ and Φ be quantifier-free formulas of $\mathcal{Q}^u$ having at most the index i free. Then for any given n there are verity sentential sentences C and A (i.e., both free of '$\dashv$') such that

$$\bigwedge_{i=0}^{n}(\Psi \dashv \Phi) \equiv_u C \dashv A.$$

This is an immediate consequence of SPL, p. 253 last paragraph of §5.6, referred to above in §1.4. Note that the formula on the left is u-equivalent to a formula, namely the one on the right, having only one occurence of the suppositional symbol although Ψ and Φ are unrestricted in the number of such occurrences.

Let $\mathbf{1}$ be a formula, e.g., $A_0 \vee \neg A_0$, whose value in any model is 1. We have

EXAMPLE 3. If Ψ is a closed formula of $\mathcal{Q}^u$ and is quantifier-free then

$$\Psi \dashv \mathbf{1} \equiv_u \Psi.$$

This is evident from (2) above since

$$U(\Psi \dashv \mathbf{1}) = \max\{\min\{1, U(\Psi)\}, \min\{0, u\}\}$$
$$= \max\{\min\{U(\Psi)\}, 0\}$$
$$= \max\{\min\{U(\Psi)\}\}$$
$$= U(\Psi).$$

With the introduction in Chapter 4 of probability semantics for it our interest in quantifier logic with the suppositional will be resumed.

CHAPTER 3

PROBABILITY SEMANTICS FOR ON LOGIC

§3.1. Probability functions on ON languages

In Chapter 1 probability logic is presented as a generalization of sentential verity logic in which, instead of just 0 and 1 as semantic values, values
in the unit interval $[0, 1]$ are allowed. Basic to the development was the
notion of a probability function. Here in this chapter we shall be extending
the notion of probability from a potential infinite sentential language $\mathcal{S}$ to
a denumerable ON language $\mathcal{Q}$ (with quantifiers described in Chapter 2),
likewise by means of a probability function on the language.

In place of having (semantic) truth-functional validity implying probability 1 (P1 of §1.2) here we employ provability in an appropriate axiomatic
formulation, the extended '⊢' of §2.2. In addition to this change in P1, we
find it convenient to replace the P2, P3 of §1.2 by the properties (b) and
(d) listed there. The addition now of P4 to these and taking the arguments
for P to be closed formulas of $\mathcal{Q}$ will extend the concept of probability
to closed quantifier formulas. (Since this is an extension of the concept
of probability from sentential to quantification logic to which P1–P3 still
apply we shall continue to use the same symbol 'P' though now applying
to the so enlarged language.)

Let $\mathcal{Q}$ be an ON language and $\mathcal{Q}^{cl}$ its set of closed formulas. A *probability
function* (for $\mathcal{Q}$) is a function $P \colon \mathcal{Q}^{cl} \to [0, 1]$ such that, for any (neccessarily
closed) formulas ϕ, ψ, and both $(\bigvee i_1) \cdots (\bigvee i_r)\chi$ and $(\bigwedge i_1) \cdots (\bigwedge i_r)\chi$ for
arbitrary r are in $\mathcal{Q}^{cl}$, the following hold:

P1. If $\vdash \phi$, then $P(\phi) = 1$.

P2. $P(\phi) + P(\neg\phi) = 1$.

P3. $P(\phi \vee \psi) = P(\phi) + P(\psi) - P(\phi\psi)$.

P4. $P((\bigvee i_1) \cdots (\bigvee i_r)\chi) = \lim_{n \to \infty} P(\bigvee_{i_1=0}^{n} \cdots \bigvee_{i_r=0}^{n} \chi)$.

$P((\bigwedge i_1) \cdots (\bigwedge i_r)\chi) = \lim_{n \to \infty} P(\bigwedge_{i_1=0}^{n} \cdots \bigwedge_{i_r=0}^{n} \chi)$.

Note that in P4 our semantics is including the notion of the limit of a sequence of real numbers, and that 'n' is being used both as a numeral variable (in the scope of P) and as the corresponding number variable (under the 'lim').[1]

Defining a probability function as a function P that satisfies P1–P4 doesn't bring one into existence—there may not be any satisfying the definition. That there are such that do is a consequence of Theorem 3.21 below, where it is shown that any probability model M for an ON language $\mathcal{Q}$ (definition below) engenders a unique probability function on $\mathcal{Q}^{cl}$. In this context, i.e., that of a probability model, P4 is providing a meaning for the probability of a quantified formula.

Assuming that there is a P that satisfies P1–P3 one readily deduces for it (P applying always to closed formulas)

(a) If $\vdash \neg(\phi\psi)$, then $P(\phi \vee \psi) = P(\phi) + P(\psi)$

(b) If $\vdash \phi \to \psi$, then $P(\phi) \leq P(\psi)$

(c) If $\vdash \phi \leftrightarrow \psi$, then $P(\phi) = P(\psi)$

(d) If $\vdash \phi \leftrightarrow \psi$, then $P(\mathcal{F}(\phi)) = P(\mathcal{F}(\psi))$,

where in (d) the formula $\mathcal{F}(\psi)$ is the result of replacing ϕ in $\mathcal{F}(\phi)$ by ψ.

Many properties of a probability function, such as "asserted" by P2–P4, are independent of any hypothesis (other than closure of the formulas that

---

[1] This 'abuse' of notation will be used on other occasions, e.g., as in the proof of (g) and in Theorem 3.11 below. We think this is preferable to, say, introducing additional notation such as bold face letters denoting numerals for their corresponding numbers in normal face type.

P applies to). Some others, readily derived, are:

(e) $P(\phi\psi) + P(\phi\neg\psi) = P(\phi)$

(f) $P(\phi \vee \psi) \leq P(\phi) + P(\psi)$

(g) $P((\bigwedge i_1)\cdots(\bigwedge i_r)\phi) = 1 - P((\bigvee i_1)\cdots(\bigvee i_r)\neg\phi)$.

As an illustration we sketch the proof of (g):

$$
\begin{aligned}
P((\bigwedge i_1)\cdots(\bigwedge i_r)\phi) &= \lim_{n\to\infty} P(\bigwedge_{i_1=0}^n \cdots \bigwedge_{i_r=0}^n \phi) \\
&= \lim_{n\to\infty} [1 - P(\bigvee_{i_1=0}^n \cdots \bigvee_{i_r=0}^n \neg\phi)] \\
&= 1 - \lim_{n\to\infty} P(\bigvee_{i_1=0}^n \cdots \bigvee_{i_r=0}^n \neg\phi) \\
&= 1 - P((\bigvee i_1)\cdots(\bigvee i_r)\neg\phi).
\end{aligned}
$$

The following significant result in quantifier probability logic, generalizing (f), is another:

THEOREM 3.11. (Countable sub-additivity) *If ϕ is a formula of $\mathcal{Q}$ with one free index i, then*

$$
P((\bigvee i)\phi) \leq \sum_{i=0}^{\infty} P(\phi).
$$

PROOF. We have ($\phi(\nu)$ abbreviating $\phi[\nu/i]$ for any numeral ν)

$$
\begin{aligned}
P(\phi(0) \vee \phi(1)) &= P(\phi(0) \vee \neg\phi(0)\phi(1)) & \text{by (c)} \\
&= P(\phi(0)) + P(\neg\phi(0)\phi(1)) & \text{by (a)} \\
&\leq P(\phi(0)) + P(\phi(1)). & \text{by (b)}
\end{aligned}
$$

Then by induction

$$
P(\bigvee_{i=0}^{\nu}\phi) \leq \sum_{i=0}^{\nu} P(\phi)
$$

so that on taking limits, and making use of P4, the result follows. $\square$

The definition of *probability model for an* ON *language $\mathcal{Q}$* is the same as that for a sentential language $\mathcal{S}$ as given in §1.2 II except for the appropriate change here of having the constituents constructed from the list $A_{\epsilon_1}, A_{\epsilon_2}, \ldots, A_{\epsilon_i}, \ldots$ instead of $A_1, \ldots, A_n, \ldots$; where for any ON language $\mathcal{Q}$ its atomic sentences are $A_{\epsilon_1}, \ldots, A_{\epsilon_i}, \ldots$, with $\epsilon_1, \ldots, \epsilon_i, \ldots$ being

a list of $\mathcal{Q}$'s numeral expressions whose values are in one-to-one corre-
spondence with the numerals $0, 1, \ldots, n, \ldots$. Let $\mathcal{Q}^0$ denote the set of $\mathcal{Q}$'s
sentential formulas constructed from this sequence (the A_{ϵ_i}) and consider
a model defined for $\mathcal{Q}^0$ using this sequence instead of the A_i of $\mathcal{S}$. With $\mathcal{Q}^0$
playing the role of $\mathcal{S}$, and with a probability model for $\mathcal{Q}$ being the same
as one for $\mathcal{Q}^0$, we can immediately take over, appropriately converted, the
extension result stated in §1.2 **II**:

THEOREM 3.12. *For each probability model M of $\mathcal{Q}$ there is a unique
function P_M^0 extending M so as to be a probability function on $\mathcal{Q}^0$, the
closed quantifier-free formulas of $\mathcal{Q}$.*

NOTE. This defines P_M^0 and its properties for M and Q^0 (Q^0 understood,
though not entirely depicted, only the superscript ^{0} on P_M^0).

Our next section generalizes this result, showing that a probability model
can be uniquely extended to be a probability function for all of ON quantifier language not just for its quantifier free formulas. This establishes that,
what were only provisional properties of a P under the assumption that it
was a function that satisfied P1–P4, will be justified, by Theorem 3.21 in
the next section, for a P_M determined by a probability model M.

That probability can be introduced so as to apply to quantifier formulas
was, apparently, first accomplished in *Gaifman* 1964 for first order predicate
logic, and for more general languages in *Scott and Krauss* 1966. The basic
ideas for a treatment of the probability of quantifications, appropriately
adapted to ON logic, were suggested to us by the Gaifman paper.

Before continuing we present a simple (prenex) example of how property
P4 of the definition of a probability function takes care of, i.e., establishes
a meaning for, a formula having both types of quantifiers.

EXAMPLE. Expanding $P((\bigwedge i)(\bigvee j)\phi(i,j))$ by P4.

$$
\begin{aligned}
P((\textstyle\bigwedge i)(\bigvee j)\phi(i,j)) &= \lim_{n\to\infty} P(\textstyle\bigwedge_{i=0}^{n}(\bigvee j)\phi(i,j)) \\
&= \lim_{n\to\infty} P[(\textstyle\bigvee j)\phi(0,j)(\bigvee j)\phi(1,j)\cdots(\bigvee j)\phi(n,j)] \\
&= \lim_{n\to\infty} P[(\textstyle\bigvee j_0)\cdots(\bigvee j_n)\Phi^n(j_0,\ldots,j_n)] \qquad (1)
\end{aligned}
$$

where in (1)

$$\Phi^n(j_0,\ldots,j_n) \text{ is } \phi(0,j_0)\cdots\phi(n,j_n).$$

Now expanding by P4 the probability on the right in (1),

$$P((\bigvee j_0) \cdots (\bigvee j_n) \Phi^n(j_0, \ldots, j_n)) = \lim_{m \to \infty} P(\bigvee_{j_0=0}^{m} \cdots \bigvee_{j_n=0}^{m} \Phi^n(j_0, \ldots, j_n)).$$

Hence with (1),

$$P((\bigwedge i)(\bigvee j)\phi(i,j)) = \lim_{n \to \infty} \lim_{m \to \infty} P(\bigvee_{j_0=0}^{m} \cdots \bigvee_{j_n=0}^{m} \Phi^n(j_0, \ldots, j_n)).$$

§3.2. Main Theorem of ON Probability Logic

This section establishes the fundamental semantic result for quantifier probability logic: a probability model for an ON language $\mathcal{Q}$, assigning a probability value to each of its atomic sentences, can be uniquely extended so as to be a probability function, i.e., a function satisfying P1–P4 of §3.1, on the closed formulas of $\mathcal{Q}$.[2] Briefly, that a probability model can be extended to be a probability function on $\mathcal{Q}^{cl}$. This is somewhat comparable, though much more complex, to the truth tables for $\neg$, $\wedge$, $\vee$ sufficing to establish, for any model for $\mathcal{S}$, a truth value for every sentence of $\mathcal{S}$.

Before attending to the proof of this main theorem (Theorem 3.21 below) we need to develop some auxiliary material.

We shall be using an operation on a formula, $\sharp$, placed as an exponent on it so designating a uniquely specified logically equivalent one *in prenex form* to be described. In this connection we adopt the convention of using 'κ' to stand for either '$\wedge$' or '$\vee$' and that along with negation they are the only connectives involved. Also we use '$(\bigwedge)$', respectively '$(\bigvee)$', to stand for an arbitrary succession of $\bigwedge$-, respectively $\bigvee$-quantifiers, referring to these as *blocks* of quantifiers. Subscripts on $\bigwedge$ or $\bigvee$ are used to distinguish

---

[2]We are indebted to Shaughan Lavine for having called our attention to the need for a result of this nature. To appreciate how much this entailed one need only look at the 2 pages needed to establish the result for ON verity logic (Theorem 2.11), and now the 10 pages needed here to establish the comparable one for ON probability logic (Theorem 3.21).

such blocks, any two in the same context assumed to have no index in common, and a prime on $\bigwedge$, or $\bigvee$, and also on its scope, indicates that its bound indices have been 'relettered' so as to be distinct from other indices in the same context. In what follows uses of Theorem 2.20 carrying over valid results from predicate to ON logic will often be used without explicit mention. The purpose of *Step* 3 of the definition (to be stated) is to eliminate spurious counting of block numbers if there should be vacuous quantifiers present. For example, if the index i occurs in α but not in χ then $(\bigwedge i)(\alpha\chi \vee \neg\alpha\chi)$, or $(\bigvee i)(\alpha\chi \vee \neg\alpha\chi)$, is truth-functionally equivalent to χ and neither $(\bigwedge i)$ nor $(\bigvee i)$ is to be included in the count of quantifiers. And, likewise, if a component of the formula can be converted to the form $\alpha\chi$, α truth-functionally valid, or to the form $\alpha \vee \chi$, α truth-functionally invalid, the α and any quantifier in the prefix whose index occurs only in α can be deleted. In what follows we shall assume that "closed prenex form" tacitly includes "with no vacuous quantifiers in the prefix".

DEFINITION OF $\phi^\sharp$ FOR AN ON FORMULA ϕ

The formula $\phi^\sharp$ logically equivalent to ϕ is obtained from ϕ by use of the following three steps used repeatedly as long as change ensues.

Step 1. By use of the rules of passage for $\neg$, move all occurrences of $\neg$ inwards until no quantifier appears within the scope of a $\neg$.

Step 2. Place a $\sharp$ on the formula and on each subformula containing a κ; then recursively use the following replacement rules working from innermost ones outwards until no further change ensues or the $\sharp$ is removed by (c) ('$\Rightarrow$' stands for 'is replaceable by' and some additional explanation follows this list):

(a) $(\theta \,\kappa\, \phi)^\sharp \Rightarrow \theta^\sharp \,\kappa\, \phi^\sharp$, $((\bigwedge)\theta)^\sharp \Rightarrow (\bigwedge)\theta^\sharp$, $((\bigvee)\theta)^\sharp \Rightarrow (\bigvee)\theta^\sharp$

(b) $((\bigwedge_1)\theta_1 \,\kappa\, (\bigwedge_2)\theta_2)^\sharp \Rightarrow (\bigwedge_1)(\bigwedge_2')(\theta_1 \,\kappa\, \theta_2')^\sharp \Rightarrow (\bigwedge_3)(\theta_1 \,\kappa\, \theta_2')^\sharp$

$((\bigvee_1)\theta_1 \,\kappa\, (\bigvee_2)\theta_2)^\sharp \Rightarrow (\bigvee_1)(\bigvee_2')(\theta_1 \,\kappa\, \theta_2')^\sharp \Rightarrow (\bigvee_3)(\theta_1 \,\kappa\, \theta_2')^\sharp$

$((\bigwedge_1)\theta_1 \,\kappa\, (\bigvee_2)\theta_2)^\sharp \Rightarrow (\bigwedge_1)(\bigvee_2')(\theta_1 \,\kappa\, \theta_2')^\sharp$

$((\bigvee_1)\theta_1 \,\kappa\, (\bigwedge_2)\theta_2)^\sharp \Rightarrow (\bigvee_1)(\bigwedge_2')(\theta_1 \,\kappa\, \theta_2')^\sharp$

(c) $\phi^\sharp \Rightarrow \phi$, if ϕ is quantifier-free.

To combine similar cases for this step we allow the possibility of an indicated block being empty; for example, referring to the first item in row 1 of (b), if $(\bigwedge_2)$ were to be 'empty' then the item is to be considered as

$$((\textstyle\bigwedge_1)\theta_1 \; \kappa \; \theta_2)^\sharp \Rightarrow (\textstyle\bigwedge_1)(\theta_1 \; \kappa \; \theta_2)^\sharp.$$

Also in this line of (b), $(\bigwedge_1)(\bigwedge_2')$, being a block of $\bigwedge$-quantifiers, is called '$(\bigwedge_3)$'. Blocks are to be moved only in their entirety. Of a conjoined (by κ) pair, each starting with a block, it is the leftmost that is to come out first.

Step 3. (Deletion of vacuous quantifiers) If the end result of Steps 1 and 2 is of the form $(\Pi)\theta$, with (Π) the prefix and θ the matrix, replace θ by a truth-functionally equivalent formula θ' having the fewest number of indices (if not already the fewest) and delete from (Π) any quantifiers corresponding to indices not present in θ' so as to obtain the prefix (Π').

The *block number* of a formula ϕ is the number of alternating (between universal and existential) blocks of quantifiers in $\phi^\sharp$'s prefix.

Since only rules of passage and removal of vacuous quantifiers are involved in these steps, so resulting in an equivalent formula, we have

LEMMA 1. *For any formulas ϕ and ψ,*

(a) $\vdash \phi \leftrightarrow \phi^\sharp$.

(b) $\vdash \phi \leftrightarrow \psi$ *if and only if* $\vdash \phi^\sharp \leftrightarrow \psi^\sharp$.

(c) $\vdash (\neg\phi)^\sharp \leftrightarrow \neg\phi^\sharp$.

Part (c) of this Lemma is obtained by combining these two of (a)'s immediate consequences:

$$(\text{i}) \; \vdash \neg\phi \leftrightarrow \neg\phi^\sharp,$$

obtained by negating both sides of the equivalence (a), and

$$(\text{ii}) \; (\neg\phi) \leftrightarrow (\neg\phi)^\sharp,$$

obtained by replacing both occurrences of 'ϕ' in (a) by '$\neg\phi$'.

To eliminate ambiguity in the case of an unparenthesized succession of $\wedge$ or $\vee$, we specify that association to the left is to be understood. For example, the formula

$$\bigvee_{i=0}^{2}(\wedge j)(\vee k)\phi(i,j,k) \tag{1}$$

is to be understood as

$$((\wedge j)(\vee k)\phi(0,j,k) \vee (\wedge j)(\vee k)(\phi(1,j,k)) \vee (\wedge j)(\vee k)\phi(2,j,k).$$

We abbreviate this (1) to (employing brackets to enhance readability)

$$[(\wedge_0)(\vee_0)\phi_0 \vee (\wedge_1)(\vee_1)\phi_1] \vee (\wedge_2)(\vee_2)\phi_2, \tag{2}$$

and now obtain its $\sharp$ as an example illustrating that for the general case.

Assuming the ϕ_i here to be quantifier-free so that by Step 2(c) $(\phi_i)^{\sharp}$ is ϕ_i, that all the bound variables are distinct and hence $((\wedge_i)(\vee_i)\phi_i)^{\sharp}$ is $(\wedge_i)(\vee_i)\phi_i$, then the $\sharp$ of (2) is obtained in the following five steps making use of the definition of $\sharp$, the first one by its (a) and the others by its (b):

$$([(\wedge_0)(\vee_0)\phi_0 \vee (\wedge_1)(\vee_1)\phi_1]^{\sharp} \vee (\wedge_2)(\vee_2)\phi_2)^{\sharp}$$
$$((\wedge_0)(\wedge_1')[(\vee_0)\phi_0 \vee (\vee_1)\phi_1']^{\sharp} \vee (\wedge_2)(\vee_2)\phi_2)^{\sharp}$$
$$((\wedge_0)(\wedge_1')(\vee_0)(\vee_1')(\phi_0 \vee \phi_1'') \vee (\wedge_2)(\vee_2)\phi_2)^{\sharp}$$
$$(\wedge_0)(\wedge_1')(\wedge_2')((\vee_0)(\vee_1')(\phi_0 \vee \phi_1'') \vee (\vee_2)\phi_2')^{\sharp}$$
$$(\wedge_0)(\wedge_1')(\wedge_2')(\vee_0)(\vee_1')(\vee_2')(\phi_0 \vee \phi_1'' \vee \phi_2''). \tag{3}$$

Note that each of the three $\vee$-terms in formula (1), as well as its $\sharp$ as shown in (3) has block number 2. This example, i.e., (3) as the $\sharp$ of (1), is an illustration of the following LEMMA 2 for the special case of the Lemma's r being 1, it's $b-1$ being 2 and it's n being 2 and with

$$(\bigvee_{i=0}^{2}(\wedge j)(\vee k)\phi(i,j,k))^{\sharp}$$

an instance of (4) in the following Lemma whose proof establishes the validity of this conversion example for the general case:

LEMMA 2. *Suppose ϕ is a formula in prenex form whose prefix has b alternating blocks of quantifiers beginning with an $\bigwedge$-block, i.e., ϕ with its initial block displayed, is of the form*

$$(\textstyle\bigwedge i_1) \cdots (\bigwedge i_r)\theta,$$

θ being the remaining portion of the prenex formula. Then for any numeral n, the formula

$$(\textstyle\bigwedge_{i_1=0}^{n} \cdots \bigwedge_{i_r=0}^{n} \theta)^{\sharp} \tag{4}$$

designates a prenex formula with $b - 1$ blocks of quantifiers.

A similar statement holds if ϕ begins with an $\bigvee$-block, i.e., if prenex $(\bigvee)\theta$ has b alternating blocks of quantifiers then $(\bigvee_{i_1=0}^{n} \cdots \bigvee_{i_r=0}^{n} \theta)^{\sharp}$ is prenex with $b - 1$ blocks of quantifiers.

PROOF. In (4) the formula to which the $\sharp$ operator is applied is a conjunction of $(n+1)^r$ conjunctions each of which begins with the same (apart from alphabetical changes of indices) $b - 1$ blocks that θ does. The operation then brings out quantifier blocks to prefix position in corresponding groups, i.e., each first block, then each second, etc. The size of the blocks thus resulting is increased, but not the number of blocks, which is the same as that of θ, namely $b - 1$. □

Let $\mathcal{Q}^b$, for each $b = 0, 1, 2 \ldots$, be the set of closed formulas having block number b or less. (Note: Hence $\mathcal{Q}^b$ includes $\mathcal{Q}^a$ if $a < b$ and $\mathcal{Q}^0$ is as defined in Theorem 3.12.) Despite its lengthy proof the content of the following Lemma, simply stated, is that a probability function defined on $\mathcal{Q}^0$ has at most one extension which is a probability function on $\mathcal{Q}^b$.

LEMMA 3. *Let P_M^0 be as described in Theorem 3.12. For any $b > 0$, if P_1 and P_2 are probability functions on $\mathcal{Q}^b$ agreeing with P_M^0 on $\mathcal{Q}^0$ then they are identical on $\mathcal{Q}^b$.*

PROOF. Let P_1 and P_2 be functions as described in the hypothesis. In order to show that they are identical on $\mathcal{Q}^b$ it suffices to show that for any new ϕ (i.e., one whose $\phi^{\sharp}$ is not already in $\mathcal{Q}^a$, $a < b$) one has $P_1(\pi) = P_2(\pi)$, where π is a prenex form of ϕ; for a closed formula is logically equivalent to any of its prenex forms and, both P_1 and P_2 being probability functions on $\mathcal{Q}^b$, they each satisfy (c) of §3.1, i.e., $P_1(\phi) = P_1(\pi)$ and $P_2(\phi) = P_2(\pi)$.

So that if also $P_1(\pi) = P_2(\pi)$ then $P_1(\phi) = P_2(\phi)$. As our prenex form for a ϕ in $\mathcal{Q}^b$ we select $\phi^\sharp$. (By Lemma 1 (a), $\vdash \phi \leftrightarrow \phi^\sharp$.)

The proof that for any b the functions P_1 and P_2 are identical on $\mathcal{Q}^b$ is by induction on b. The induction need only start with $b = 1$ since by definition the functions agree with P_M^0 on $\mathcal{Q}^0$.

BASIS. $b = 1$

Here $\phi^\sharp$ is either (a$_1$) $(\bigvee i_1) \cdots (\bigvee i_r)\theta$ or (a$_2$) $(\bigwedge i_1) \cdots (\bigwedge i_r)\theta$, θ being quantifier-free with $i_1, \ldots, i_r$ its indices.

Case (a$_1$). $\phi^\sharp$ is $(\bigvee i_1) \cdots (\bigvee i_r)\theta$.

Since by Lemma 2 for any n, $\bigvee_{i_1=0}^n \cdots \bigvee_{i_r=0}^n \theta$—being quantifier-free— is in $\mathcal{Q}^0$ and by hypothesis P_1 and P_2 coincide on $\mathcal{Q}^0$, we have

$$P_1(\bigvee_{i_1=0}^n \cdots \bigvee_{i_r=0}^n \theta) = P_2(\bigvee_{i_1=0}^n \cdots \bigvee_{i_r=0}^n \theta),$$

so that on taking limits one has

$$\lim_{n \to \infty} P_1(\bigvee_{i_1=0}^n \cdots \bigvee_{i_r=0}^n \theta) = \lim_{n \to \infty} P_2(\bigvee_{i_1=0}^n \cdots \bigvee_{i_r=0}^n \theta). \qquad (5)$$

Since P_1 and P_2 are probability functions on $\mathcal{Q}^0$ the limits in (5) exist; and inasmuch as being probabilities the P_1 and P_2 values in (5) are bounded above by 1 and monotonically non-decreasing with increasing n. Hence by P4, §3.1,

$$P_1((\bigvee i_1) \cdots (\bigvee i_r)\theta) = P_2((\bigvee i_1) \cdots (\bigvee i_r)\theta).$$

Case (a$_2$). $\phi^\sharp$ is $(\bigwedge i_1) \cdots (\bigwedge i_r)\theta$.

Proof here is similar to that of Case (a$_1$) with the values of P_1 and P_2 in the equation corresponding to (5) now bounded below by 0 and monotonically non-increasing. Hence the limits exist and are equal.

INDUCTION STEP. Assume as hypothesis of induction that if P_1 and P_2 are probability functions extending P_M^0 from $\mathcal{Q}^0$ to $\mathcal{Q}^{b-1}$ then they are identical. We need to show that the same statement is true with b in place of $b - 1$. It suffices to consider, as argument for the P's, a ϕ in $\mathcal{Q}^b$ but not in $\mathcal{Q}^{b-1}$ since by hypothesis P_1 and P_2 are identical on $\mathcal{Q}^{b-1}$. This ϕ has block number b. We have two cases according as $\phi^\sharp$ is of the form $(\bigvee i_1) \cdots (\bigvee i_r)\theta^\sharp$ or $(\bigwedge i_1) \cdots (\bigwedge i_r)\theta^\sharp$.

Case (b$_1$). $\phi^\sharp$ is $(\bigvee i_1) \cdots (\bigvee i_r)\theta^\sharp$, with $\theta^\sharp$ having block number $b - 1$.

Then similarly to the argument in the BASIS step one has

$$P_1((\bigvee_{i_1=0}^{n} \cdots \bigvee_{i_r=0}^{n} \theta)^{\sharp}) = P_2((\bigvee_{i_1=0}^{n} \cdots \bigvee_{i_r=0}^{n} \theta)^{\sharp})$$

and then

$$\lim_{n \to \infty} P_1((\bigvee_{i_1=0}^{n} \cdots \bigvee_{i_r=0}^{n} \theta)^{\sharp}) = \lim_{n \to \infty} P_2((\bigvee_{i_1=0}^{n} \cdots \bigvee_{i_r=0}^{n} \theta)^{\sharp}),$$

so that with the limits existing, this is

$$P_1(((\bigvee i_1) \cdots (\bigvee i_r)\theta)^{\sharp}) = P_2(((\bigvee i_1) \cdots (\bigvee i_r)\theta)^{\sharp}),$$

i.e., $P_1(\phi^{\sharp}) = P_2(\phi^{\sharp})$.

Case (b$_2$). Similarly as in the BASIS for a $\phi^{\sharp}$ beginning, however, with a $\bigwedge$-block. $\qquad\qquad\square$

LEMMA 4. *Let M be a probability model for $\mathcal{Q}$. Then there is a probability function P_M^b such that (i) P_M^b is defined for elements of $\mathcal{Q}^b$ and (ii) agrees with P_M^a for $a < b$; i.e., $P_M^b(\phi) = P_M^a(\phi)$ when ϕ is also in $\mathcal{Q}^a$.*

PROOF. By induction on b.

BASIS. $b = 0$.

For part (i) of the Lemma, that P_M^0 is defined for elements of $\mathcal{Q}^0$, follows from Theorem 3.12. As for part (ii), it is vacuously true since if $b = 0$ there is no $a < b$.

INDUCTION STEP. Assume as hypothesis of induction that P_M^{b-1} is a probability function on $\mathcal{Q}^{b-1}$ and when ϕ is in $\mathcal{Q}^a$, $a < b-1$, then $P_M^{b-1}(\phi) = P_M^a(\phi)$.

Let ϕ be a member of $\mathcal{Q}^b$.

PART I. Definition of P_M^b, $b > 0$.

Case Ia. $\phi^{\sharp}$ is of the form $(\bigvee i_1) \cdots (\bigvee i_r)\theta^{\sharp}$.

Then the following is proposed as the definition of P_M^b (for this Case 1a).

$$P_M^b(\phi) := \begin{cases} P_M^{b-1}(\phi), & \text{if } \phi \in \mathcal{Q}^{b-1} \\ \lim_{n \to \infty} P_M^{b-1}((\bigvee_{i_1=0}^{n} \cdots \bigvee_{i_r=0}^{n} \theta)^{\sharp}), & \phi \notin \mathcal{Q}^{b-1} \end{cases} \qquad (6)$$

By Lemma 2 the formula $(\bigvee_{i_1=0}^{n} \cdots \bigvee_{i_r=0}^{n} \theta)^{\sharp}$ is, for any n, of block number $b-1$. Hence by induction hypothesis $(\bigvee_{i_1=0}^{n} \cdots \bigvee_{i_r=0}^{n} \theta)^{\sharp}$ has a defined P_M^{b-1}

value. But to have (6) define one for P_M^b requires furthermore that the limit exists. To see that this is the case note that the logical validity

$$\vdash (\bigvee_{i_1=0}^{n} \cdots \bigvee_{i_r=0}^{n} \theta)^{\sharp} \rightarrow (\bigvee_{i_1=0}^{n+m} \cdots \bigvee_{i_r=0}^{n+m} \theta)^{\sharp}$$

holds for all n and m. Then, P_M^{b-1} being by hypothesis of induction a probability function on $\mathcal{Q}^{b-1}$, this logical validity leads to

$$P_M^{b-1}((\bigvee_{i_1=0}^{n} \cdots \bigvee_{i_r=0}^{n} \theta)^{\sharp}) \leq P_M^{b-1}((\bigvee_{i_1=0}^{n+m} \cdots \bigvee_{i_r=0}^{n+m} \theta)^{\sharp}).$$

Thus the P_M^{b-1} values of $(\bigvee_{i_1=0}^{n} \cdots \bigvee_{i_r=0}^{n} \theta)^{\sharp}$ on the right in (6) are monotonically non-decreasing with increasing n and, as probabilities, they are bounded above by 1. Hence the limit exists, so that for this case P_M^b is defined as a function on $\mathcal{Q}^b$.

Case Ib. $\phi^{\sharp}$ is of the form $(\bigwedge i_1) \cdots (\bigwedge i_r) \theta^{\sharp}$.

The proof here is similar to that of Case Ia with the difference that the P_M^{b-1} values of $(\bigwedge_{i_1=0}^{n} \cdots \bigwedge_{i_r=0}^{n} \theta)^{\sharp}$ are monotonically non-increasing with increasing n and are bounded below by 0.

This establishes that P_M^b proposed in (6) is defined on $\mathcal{Q}^b$ and is such that for ϕ in $\mathcal{Q}^a$, $a < b$, $P_M^b(\phi) = P_M^a(\phi)$. Hence, $P_M^b((\bigvee_{i_1=0}^{n} \cdots \bigvee_{i_r=0}^{n} \theta)^{\sharp}) = P_M^{b-1}((\bigvee_{i_1=0}^{n} \cdots \bigvee_{i_r=0}^{n} \theta)^{\sharp})$ since here the argument for P_M^b is in $\mathcal{Q}^{b-1}$. Similarly for $\bigwedge$ in place of $\bigvee$.

Having defined P_M^b in Part I, to complete the result we need to show

PART II. P_M^b, as defined, is a *probability* function.

We are still operating under the hypothesis of induction that P_M^{b-1} is a probability function on $\mathcal{Q}^{b-1}$ and that if ϕ is in $\mathcal{Q}^a$, $a < b - 1$, then $P_M^{b-1}(\phi) = P_M^a(\phi)$. Our task now is to show that the function P_M^b is a probability function, i.e., that it satisfies the postulated properties P1–P4 of §3.1.

CASE OF P1: If $\vdash \phi$ then $P(|\phi|) = 1$, $|\phi|$ being a closure of ϕ.

To show this let $\sigma_1, \sigma_2, \ldots, \sigma_p$ (σ_p being ϕ) be a formal deduction (as defined in §2.2) of ϕ, each σ_i ($i = 1, \ldots, p$) having a closure in $\mathcal{Q}^b$. Using induction on p, the length of a deduction, we show that the closure of σ_p has P_M^b value 1.

BASIS. $p = 1$.

Here σ_1 is quantifier-free and truth-functionally valid with free indices, say, $i_1, \ldots, i_r$. (See the remarks immediately following presentation of the axiom system with rules of inference **R1** to **R4** in §2.2.) Then by PART I Case Ib, just shown, with $|\sigma_1|$ being in $\mathcal{Q}^b$,

$$P_M^b((\bigwedge i_1)\cdots(\bigwedge i_r))\sigma_1) = \lim_{n\to\infty} P_M^{b-1}((\bigwedge_{i_1=0}^n \cdots \bigwedge_{i_r=0}^n \sigma_1)^\sharp)$$

$$= \lim_{n\to\infty} P_M^{b-1}(\bigwedge_{i_1=0}^n \cdots \bigwedge_{i_r=0}^n \sigma_1)$$

$$= 1,$$

where the second equation follows from the first in that σ_1 is quantifier-free, and the third from the second in that for each n the conjunction consists of conjuncts each of which is truth-functionally valid, inasmuch as $\sigma_1(i_1,\ldots,i_r)$ is truth-functionally valid.

INDUCTION STEP (for CASE OF P1). As hypothesis of induction assume that the closure of σ_i ($i = 1,\ldots,p-1$) has P_M^b value 1, and consider σ_p.

Subcase σ_p (i). σ_p comes from $\sigma_j(j < p)$ by use of **R1** (to $\bigwedge$-*introducing* i_s so σ_p is $(\bigwedge i_s)\sigma_j$).

Since here a closure of σ_p is a closure of σ_j and by hypothesis of induction that of σ_j has P_M^b value 1, then so is that of σ_p as it differs from σ_j only in having prefixed a $\bigwedge$-quantifier for one of σ_j's free indices.

Subcase σ_p (ii). σ_p comes from σ_j ($j < p$) by **R2** ($\bigvee$-*introduction*).

Let σ_p be $(\bigvee i_s)\sigma_j$ and $(\bigwedge j)$ the block of $\bigwedge$-quantifiers for all free indices of σ_j other than i_s, so that $|\sigma_j|$ is $(\bigwedge j)(\bigwedge i)\sigma_j$ and $|\sigma_p|$ is $(\bigwedge j)(\bigvee i_s)\sigma_j$. Let $\bigwedge_j^{(n)}$ be the corresponding succession of finite quantifiers for $(\bigwedge j)$. Then we have

$$\vdash (\bigwedge_j^{(n)}(\bigwedge i_s)\sigma_j)^\sharp \to (\bigwedge_j^{(n)}(\bigvee i_s)\sigma_j)^\sharp$$

with both antecedent and consequent of the conditional being in $\mathcal{Q}^{b-1}$ since both $(\bigwedge j)(\bigwedge i_s)\sigma_j$ and $(\bigwedge j)(\bigvee i_s)\sigma_j$ are in $\mathcal{Q}^b$. Since by inductive hypothesis P_M^{b-1} is a probability function on $\mathcal{Q}^{b-1}$ we have from the logical validity

$$\vdash (\bigwedge_j^{(m)}(\bigwedge i_s)\sigma_j)^\sharp \to (\bigwedge_j^{(m)}(\bigvee i_s)\sigma_j)^\sharp$$

that

$$P_M^{b-1}((\bigwedge_j^{(n)}(\bigwedge i_s)\sigma_j)^\sharp) \le P_M^{b-1}((\bigwedge_j^{(n)}(\bigvee i_s)\sigma_j)^\sharp)$$

so that

$$\lim_{n\to\infty} P_M^{b-1}((\bigwedge_j^{(n)}(\bigwedge i_s)\sigma_j)^\sharp) \le \lim_{n\to\infty} P_M^{b-1}((\bigwedge_j^{(n)}(\bigvee i_s)\sigma_j)^\sharp) \le 1,$$

i.e.,

$$P_M^b(|\sigma_j|) \le P_M^b(|\sigma_p|) \le 1.$$

Since by hypothesis of induction (for CASE OF P1) $P_M^b(|\sigma_j|) = 1$ then so also $P_M^b(|\sigma_p|) = 1$.

Subcase σ_p (iii). σ_j comes from σ_j $(j < p)$ by **R3** *(passage)*.

In this case the initial block of finitized closures of σ_p and σ_j would be logically equivalent—not just one logically implying the other—and both in $\mathcal{Q}^{b-1}$ so that when similarly treated as in Subcase σ_p (ii) their P_M^b values would come out to be equal. Hence $P_M^b(|\sigma_p|) = 1$ then so also $P_M^b(|\sigma_p|) = 1$.

Subcase σ_p (iv). σ_p comes from σ_j $(j < p)$ by a use of **R4** *(simplification)*.

The argument here is similar to that of Subcase σ_p (iii). This completes CASE OF P1.

CASE OF P2: $P(\phi) + P(\neg\phi) = 1$.

Here we need to show that, for ϕ in $\mathcal{Q}^b$,

$$P_M^b(\phi) + P_M^b(\neg\phi) = 1. \tag{7}$$

This equation holds for $b = 0$ since P_M^0 is a probability function on $\mathcal{Q}^0$, and so constitutes the BASIS for a proof by induction.

INDUCTION STEP (for CASE OF P2). As hypothesis of induction assume that (7) holds for P_M^{b-1} and consider a ϕ in $\mathcal{Q}^b$.

Subcase P2 (i). $\phi^\sharp$ is of the form $(\bigvee i_1)\cdots(\bigvee i_r)\theta^\sharp$.

Then since, by Lemma 2, $(\bigvee_{i_1=0}^n \cdots \bigvee_{i_r=0}^n \theta)^\sharp$ is in $\mathcal{Q}^{b-1}$ we have by hypothesis of induction (P_M^{b-1} being a probability function on $\mathcal{Q}^{b-1}$),

$$P_M^{b-1}((\bigvee_{i_1=0}^n \cdots \bigvee_{i_r=0}^n \theta)^\sharp) + P_M^{b-1}(\neg(\bigvee_{i_1=0}^n \cdots \bigvee_{i_r=0}^n \theta)^\sharp) = 1.$$

Then on taking limits $(n \to \infty)$ and using PART I of the Induction Step (see (6)) we have

$$P_M^b((\bigvee i_1)\cdots(\bigvee i_r)\theta) + P_M^b(\neg(\bigvee i_1)\cdots(\bigvee i_r)\theta) = 1.$$

Subcase P2 (ii). $\phi^\sharp$ is of the form $(\bigwedge i_1)\cdots(\bigwedge i_r)\theta^\sharp$.

This is similar to Subcase P2 (i).

CASE OF P3: $P(\phi \vee \psi) = P(\phi) + P(\psi) - P(\phi \wedge \psi)$.

We need to show

$$P_M^b(\phi \vee \psi) = P_M^b(\phi) + P_M^b(\psi) - P_M^b(\phi \wedge \psi), \tag{8}$$

with ϕ, ψ, $\phi \vee \psi$, $\phi \wedge \psi$ in $\mathcal{Q}^b$. (It suffices to have b being the maximum block number of the four formulas, as this would imply that the others are in $\mathcal{Q}^b$.)

We consider the following four possibilities for the initial quantifier blocks of $\phi^\sharp$ and $\psi^\sharp$:

$$\text{(a)} \quad (\bigvee_1),\ (\bigvee_2) \qquad \text{(b)} \quad (\bigwedge_1),\ (\bigwedge_2)$$
$$\text{(c)} \quad (\bigvee_1),\ (\bigwedge_2) \qquad \text{(d)} \quad (\bigwedge_1),\ (\bigvee_2)$$

with corresponding finitized versions: $\bigvee_1^{(n)}$, $\bigvee_2^{(n)}$, etc.

Subcase P3 (a). For this subcase assuming $\phi \vee \psi$ is $(\bigvee_1)\theta \vee (\bigvee_2)\chi$, which is logically equivalent to $(\bigvee_1)(\bigvee_2)(\theta \vee \chi)$—and similarly for $\phi \wedge \psi$—we have, on using PART I (Defn. of P_M^b)

$$P_M^b(\phi \vee \psi) = \lim_{n \to \infty} P_M^{b-1}(\bigvee_1^{(n)}\bigvee_2^{(n)}(\theta \vee \chi)^\sharp)$$
$$P_M^b(\phi) = \lim_{n \to \infty} P_M^{b-1}(\bigvee_1^{(n)}\theta^\sharp)$$
$$P_M^b(\psi) = \lim_{n \to \infty} P_M^{b-1}(\bigvee_2^{(n)}\chi^\sharp)$$
$$P_M^b(\phi \wedge \psi) = \lim_{n \to \infty} P_M^{b-1}(\bigvee_1^{(n)}\bigvee_2^{(n)}(\theta \wedge \chi)^\sharp).$$

The proof of (8) then follows the pattern of CASE P2 with obvious changes. Similarly for Subcases P3 (b), (c) and (d). Finally, two cases of P4:

CASE OF P4 (a). $P((\bigvee i_1) \cdots (\bigvee i_r)\phi) = \lim_{n \to \infty} P(\bigvee_{i_1=0}^n \cdots \bigvee_{i_r=0}^n \phi)$.

We have to show

$$P_M^b((\bigvee i_1) \cdots (\bigvee i_r)\phi) = \lim_{n \to \infty} P_M^b(\bigvee_{i_1=0}^n \cdots \bigvee_{i_r=0}^n \phi)$$

or, since logically equivalent formulas have equal probability function values, on replacing the formulas by prenex forms, to show

$$P_M^b((\bigvee i_1) \cdots (\bigvee i_r)\phi^\sharp) = \lim_{n \to \infty} P_M^b(\bigvee_{i_1=0}^n \cdots \bigvee_{i_r=0}^n \phi^\sharp). \qquad (9)$$

Now by the PART I definition of P_M^b, $P_M^b(\theta) = P_M^{b-1}(\theta)$ if θ is in $\mathcal{Q}^{b-1}$, and since $(\bigvee_{i_1=0}^n \cdots \bigvee_{i_r=0}^n \phi)^\sharp$ is in $\mathcal{Q}^{b-1}$ then (9) becomes

$$P_M^b((\bigvee i_1) \cdots (\bigvee i_r)\phi^\sharp) = \lim_{n \to \infty} P_M^{b-1}((\bigvee_{i_1=0}^n \cdots \bigvee_{i_r=0}^n \phi)^\sharp)$$

which is the case, being a restatement of the definition of P_M^b (see (6)).

CASE OF P4 (b). $P((\bigwedge i_1) \cdots (\bigwedge i_r)\phi) = \lim_{n \to \infty} P(\bigwedge_{i_1=0}^n \cdots \bigwedge_{i_r=0}^n \phi)$.

The proof follows the pattern of CASE OF P4 (a).

This completes the proof of Lemma 4.   □

THEOREM 3.21. (MAIN THEOREM) *To each probability model M there is a unique probability function P_M, i.e., a function satisfying P_1–P_4, which extends M to all of Q^{cl}, the set of closed formulas of Q.*

PROOF. Viewing a function as a set of ordered pairs, (argument, function value), we define P_M as the union of the sets P_M^0, $P_M^1 \setminus P_M^0$, $P_M^2 \setminus P_M^1$, etc., where '$\setminus$' is set difference. Formally,

$$P_M := P_M^0 \cup \bigcup_{b=1}^{\infty} [P_M^b \setminus P_M^{b-1}].$$

Thus if ϕ is in Q^0 then $P_M(\phi) = P_M^0(\phi)$ since for any such ϕ, the ordered pairs $(\phi, P_M^0(\phi))$ and $(\phi, P_M^b(\phi))$ are the same for any b. And, more generally, if ϕ is in Q^b then $P_M(\phi) = P_M^b(\phi)$ since for a ϕ in Q^b, $(\phi, P_M^b(\phi)) = (\phi, P_M^c(\phi))$ for any $c \geq b$.

Having defined P_M we turn to establishing that it is a probability function. That this is the case follows from the fact that any finite collection of closed formulas can be viewed as a member of Q^b, b being the maximum block number of the collection. For this collection P_M behaves as P_M^b, which, by Lemma 4 is a probability function.   □

The theorem just presented guarantees that there is a unique probability function—and hence a probability value for each sentence of the language with appropriate properties—once a probability model is provided. For finite stochastic problems in standard (set-theoretic) probability theory one makes assignments, rather, to the elementary events (or, outcomes of a trial). Assignments to constituents are nevertheless determined in that it is assumed that in a trial one and only one of the outcomes is always the case (see SPL, §4.3). For infinite stochastic situations this parallelism breaks down making, as we shall see, for a difference. Yet in ON languages one can still make assignments to atomic sentences—so long as the values assigned are in $[0, 1]$—though there need not be a unique probability function with these values. We state this formally:

THEOREM 3.22. *For any ON language Q and for any sequence of values $p_1, \ldots, p_n, \ldots$ all of which are in [0, 1], an assignment $P(A_{\epsilon_i}) = p_i$ ($i = 1, \ldots, n, \ldots$) to the atomic sentences of Q is compatible with P being a probability function on Q^{cl}.*

PROOF. (Informal) Consider a square of unit area and each A_{ϵ_i} as a subregion with area p_i. No matter how these A_{ϵ_i} subregions are arranged in the square there will be determined, for any n, constituents K_i ($i = 1, \ldots, 2^n$) on the atomic sentences $A_{\epsilon_1}, \ldots, A_{\epsilon_n}$. Then define a probability model M with, for every K_i, $M(K_i)$ equal to the area corresponding to K_i. By Theorem 3.21, M determines a probability function. $\square$

If it is desired to *assign* probability values to formulas $\phi_1, \ldots, \phi_m$ then the situation is more complicated as it isn't always possible. The following theorem gives, for *sentential* formulas, conditions under which such an assignment can be made.

THEOREM 3.23. *Let $\phi_1, \ldots, \phi_m$ be closed sentential formulas of an ON language, the formulas altogether involving n atomic sentences. Let p_i ($i = 1, \ldots, m$) be values in $[0, 1]$. Then there is a probability function P such that $P(\phi_i) = p_i$ ($i = 1, \ldots, m$) IF the following linear equation-inequation system in variables $k_1, \ldots, k_{2^n}$ HAS A SOLUTION*

$$a_{11}k_1 + a_{12}k_2 + \cdots + a_{12^n}k_{2^n} = p_1$$
$$a_{21}k_1 + a_{22}k_2 + \cdots + a_{22^n}k_{2^n} = p_2$$
$$\vdots \qquad \vdots \qquad\qquad \vdots \qquad \vdots$$
$$a_{m1}k_1 + a_{m2}k_2 + \cdots + a_{m2^n}k_{2^n} = p_m$$
$$k_1 + k_2 + \cdots + k_{2^n} = 1$$
$$k_j > 0 \quad (j = 1, \ldots, 2^n)$$

where each coefficient a_{ij} is 0 or 1, 0 if the constituent K_j ($j = 1, \ldots, 2^n$) on the n atomic sentences is absent in the disjunctive normal form expansion of ϕ_i, and 1 if it is present.

PROOF. As may be seen from a probabilistic Venn diagram the solution, if there is one, provides a probability model defining a probability function P assigning the desired values to the ϕ_i. $\square$

One more item is needed in order to have a probability logic for ON languages, namely a definition of logical consequence. Before taking up this topic (in §3.6) we interpolate some sections of mostly historical material which will provide us with useful examples.

§3.3. Borel's denumerable probability

Borel 1909 is considered by some, e.g., *Barone and Novikoff* 1978, *von Plato* 1994, p. 8, as marking the transition from classical to modern probability theory. In Borel's paper we have the first serious treatment of probability in connection with infinitely many trials and/or outcomes. In addition, Borel's paper had a philosophical aim: by the introduction of a category of probability problems between the discrete (*"probabilités discontinues"*) and the continuous (*"probabilités continues* ou *probabilités géométriques"*), a category he referred to as *"probabilités dénombrables"*, one could avoid reliance on undefinable elements necessarily present in non-denumerable sets. The question of whether Borel accomplished this aim will be discussed in §3.7.

We briefly describe and discuss from the viewpoint of our probability logic his ideas and continue the discussion in §§3.4 and 3.7. The first one considered is the example of denumerably many [independent] trials, a trial having only two possible outcomes, *success* or *failure*. These outcomes have, at the nth trial probabilities of p_n and $q_n (= 1 - p_n)$, respectively. Borel's *Problème I* asks: What is the probability that the case of success never occurs? Letting A_0 denote the probability that success never occurs he asserts that by [an extension of] the "principle of compound probabilities" [to the denumerable][3] its value is given by an infinite product, namely by

$$A_0 = (1 - p_1)(1 - p_2) \cdots (1 - p_n) \cdots . \tag{1}$$

Postponing consideration of the legitimacy of extending the principle of compound probabilities to the infinite, he discusses (1).

If the sum of our infinite series of positive terms

$$p_1 + p_2 + \cdots + p_n + \cdots \tag{2}$$

(all in $[0, 1]$) is convergent then [by a straightforward mathematical result] so is the infinite product in (1). Excluding the obvious case of a p_i equaling 1 [in which case $A_0 = 0$] the conclusion is that, in the case of convergence of (2), A_0 has a "well determined value different from 0 and 1". And if (2)

---

[3] The principle that the probability of the conjunction of independent events is equal to the product of their probabilities.

is divergent then so is the infinite product in (1), i.e., the infinite product "diverges" to 0. This value is then attributed to A_0, i.e., as his solution to *Problème 1*.

Borel contends that in the case of convergence [of (2) and so (1)] the extension to the infinite of the principle of compound probabilities needs no justification (*"va de soi"*), and also provides the definition of the desired probability. For, considering the first n trials, classical probability principles enables one to define, and in this case, calculate, the probability that success doesn't occur in these n trials. One verifies that as n increases the value varies little, not only absolutely but also relatively to this value, so that passage to the limit is entirely justified.

The case of divergence is not the same. There is, he says, a real difference between an infinitely small probability ("variable probability approaching 0") and one that is equal to 0. He likens it to the probability of choosing a rational from among the reals, the probability of which is zero, but that doesn't mean that it is impossible: "zero probability ought not to be taken [here] as impossibility". With this understanding one can then say, for the case of divergence, that $A_0 = 0$ means that as n increases the probability that success doesn't occur tends to 0.

Despite the lack of formal treatment and the imprecision of Borel's ideas we see that our conception of an infinite conjunction in probability logic, as expressed by $(\bigwedge i)\phi$, is pretty much in accord with his ideas: from our definition of a probability function we have, by P4 in §3.1, that

$$P((\bigwedge i)\phi) = \lim_{n \to \infty} P(\bigwedge_{i=0}^{n} \phi)$$

so that our probability of an infinite conjunction is the limit of the probability of an increasing conjunction which can be 0 without any of the finite conjunctions being 0. The question of extending the principle of compound probabilities to the infinite which Borel took for granted, and a treatment of his *Problème I* is discussed in our next section.

Borel's *Problème II* also involves an extension from the finite to the infinite of the classical property: *"le principe des probabilités totales"*, i.e., that the probability of at least one of a [finite] number of mutually exclusive alternatives happening is the sum of their respective probabilities. In this second problem Borel addresses himself to determining the probability of exactly k successes (and no more) in an infinite sequence of trials. We shall limit our discussion to the case of $k = 1$. He derives probabilities for success to occur at the first trial and not thereafter, for the occurrence of success only at the 2nd trial and not thereafter, and so forth. These events are mutually exclusive. The (denumerable) sum of their probabilities is the probability of success exactly once. By way of justification of this application of the principle of total probabilities, he asserts that an argument analogous to the preceding one [the principle of compound probabilities, extended] justifies it. We shall be discussing this in §3.4 and §3.7.

Borel's third problem considers the question: What is the probability that success occurs an infinite number of times? Here unlike the two preceding situations the division between the convergence and divergence of $\sum p_i$, which he thought was needed there, is indeed significant. His result is that when the series is convergent ($\sum p_i < \infty$) the probability of success occurring an infinite number of times is 0, and when divergent ($\sum p_i = \infty$) the probability is 1. This result, that there are only these two extreme values, is known in the mathematics probability literature as the Borel-Cantelli Lemma. This also we take up in our next section.

§3.4. Infinite "events" and probability functions

In this section we derive some results involving probability functions and quantifier sentences. Situations involving infinitely many trials or outcomes are treated here from our probability logic viewpoint by means of instances of a formula with one (or more) free indices, e.g., $\phi(0), \phi(1), \ldots, \phi(\nu), \ldots$, where $\phi(\nu)$ is $\phi[\nu/i]$, i being ϕ's one free index. The quantified sentence $(\bigwedge i)\phi$ expresses the occurrence of all ϕ's instances. Here "occurrence" has for us an idiomatic meaning; it refers to an indefinitely proceeding sequence of occurrences

$$\phi(0), \ \phi(0)\phi(1), \ \ldots, \ \phi(0)\phi(1)\cdots\phi(\nu),\ldots$$

each of which is a finite conjunction. The probability of such an occurrence is, by P4 §3.1, given by

$$P((\bigwedge i)\phi) = \lim_{n \to \infty} P(\bigwedge_{i=0}^{n} \phi).$$

In this connection we introduce the notion of the *serial (probabilistic) independence of all ϕ's instances*, namely, as the property expressed by

For all ν, $P(\phi(0)\phi(1)\cdots\phi(\nu)) = P(\phi(0)\phi(1)\cdots\phi(\nu-1))P(\phi(\nu))$

or, more compactly, as

For all $\nu > 0$, $P(\bigwedge_{i=0}^{\nu}\phi) = P(\bigwedge_{i=0}^{\nu-1}\phi)P(\phi(\nu))$. \hfill (1)

We abbreviate this condition (1) to $\text{Ind}(P, \phi(\nu))$. It expresses the idea of an infinite sequence of events being such that the occurrence of any one of them is probabilistically independent of the occurrence of all the preceding ones.

Serial independence is a weaker notion than that of pairwise independence. To show this we produce an example of three sentences E, F and G, which are serially independent, i.e., satisfy

$$P(EFG) = P(EF)P(G)$$
$$= P(E)P(F)P(G), \hfill (2)$$

but do not satisfy

$$P(EG) = P(E)P(G). \hfill (3)$$

We first illustrate the idea behind the construction of the example with a probabilistic Venn diagram.

In Figure 2 let the outer rectangle have area 1 (area = probability) and the three inner rectangles, labelled E, F and G, have area equal to their respective probabilities as follows. We first assign values to the eight constituents $EFG, \overline{E}FG, \ldots, \overline{E}\,\overline{F}\,\overline{G}$ so that they add up to 1 and that equations (2) are satisfied. Then a portion of E (the small notch) is removed from the part of E not in F or G, and added to the part of E in G but not in F. This change will not effect the truth of (2) since the areas of E, F, G, EF and EFG are the same as they were. However that for (3) no longer

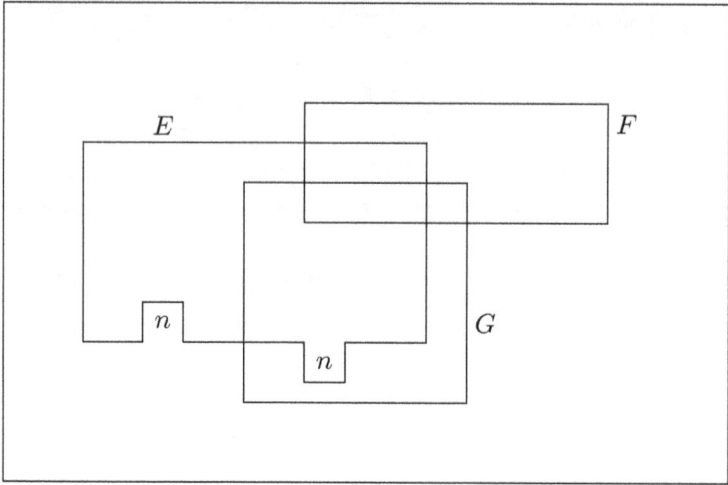

FIGURE 2.

holds since although the areas of E and G are unchanged that for EG is enlarged.

Thus (2) does not imply (3) for arbitrary E, F and G.

In terms of this notion of serial independence (defined by (1)) we state a theorem that corresponds to an extension to the infinite of the principle of compound probabilities:

THEOREM 3.41. *For any probability function P, if* $\mathrm{Ind}(P, \phi(\nu))$ *then*

$$P((\wedge i)\phi) = \prod_{i=0}^{\infty} P(\phi). \qquad (4)$$

PROOF. Assume the hypothesis. For $\nu = 1$ (using ϕ_0 as an abbreviation for $\phi(0)$, ϕ_1 for $\phi(1)$, etc.)

$$P(\phi_0 \phi_1) = P(\phi_0)P(\phi_1),$$

and for $\nu = 2$,

$$P(\phi_0 \phi_1 \phi_2) = P(\phi_0 \phi_1)P(\phi_2)$$
$$= P(\phi_0)P(\phi_1)P(\phi_2)$$

so that by induction

$$P(\bigwedge_{i=0}^{\nu}\phi) = \prod_{i=0}^{\nu} P(\phi).$$

Then on taking limits (which exist since by hypothesis P is a probability function) and using P4 the result follows.   □

In his solution of *Problème I* Borel makes no mention of independence, either of the instances of success or of failure. We assume he was taking this for granted. In our discussion we shall be using Theorem 3.41 with ϕ replaced by $\neg\phi$. This entails a change in the hypothesis from $\text{Ind}(P, \phi(\nu))$ to $\text{Ind}(P, \neg\phi(\nu))$. Since Borel makes no mention of independence it is of some interest to see that these two independence conditions are equivalent:

THEOREM 3.42. *For P a probability function, if* $\text{Ind}(P, \phi(\nu))$ *then also* $\text{Ind}(P, \neg\phi(\nu))$.

PROOF. One has (using ϕ_0 for $\phi(0)$, ϕ_1 for $\phi(1)$),

$$P(\neg\phi_0\neg\phi_1) = 1 - P(\phi_0 \vee \phi_1)$$
$$= 1 - [P(\phi_0) + P(\phi_1) - P(\phi_0\phi_1)]$$

and, by (1) for the case of $\nu = 1$,

$$= 1 - P(\phi_0) - P(\phi_1) + P(\phi_0)P(\phi_1)$$
$$= (1 - P(\phi_0))(1 - P(\phi_1))$$
$$= P(\neg\phi_0)P(\neg\phi_1).$$

With this case as the start, induction then gives

$$P(\bigwedge_{i=0}^{\nu}\neg\phi) = P(\bigwedge_{i=0}^{\nu-1}\neg\phi)P(\neg\phi(\nu))$$

which is (see (1) above) the definition of $\text{Ind}(P, \neg\phi(\nu))$.   □

We now solve Borel's *Problème I* (stated in §3.3) within ON probability logic, comparing it with his $A_0 = (1 - p_1)(1 - p_2)\cdots(1 - p_n)\cdots$ as we proceed. It serves as a nice example of how probability logic handles a non-finite situation.

In place of Borel's bland 'A_0' we have '$P((\bigwedge i)\neg B_i)$', expressing the probability of no success in the infinite sequence of trials $B_1, B_2, \ldots, B_i, \ldots$. For each i, B_i is a sentence saying that *success* is the outcome on that trial. In place of Borel's $(1-p_1)(1-p_2)\cdots(1-p_n)\cdots$ we have, $P(B_i)$ replacing p_i,

$$\prod_{i=1}^{\infty}(1-P(B_i)).$$

Then, by Theorem 3.41, if P is a probability function and $\mathrm{Ind}(P,\neg B_i)$—or equivalently by Theorem 3.42, if $\mathrm{Ind}(P, B_i)$—we have

$$P((\textstyle\bigwedge i)\neg B_i) = \prod_{i=1}^{\infty}(1-P(B_i)).$$

Thus to obtain in probability logic what corresponds to Borel's solution we need to specify a probability function P for which $\mathrm{Ind}(P, B_i)$ and $P(B_i) = p_i$, $0 < p_i < 1$ $(i = 1,2,\dots)$. Theorem 3.21 is used to obtain this P by specifying an appropriate model M as follows.

Let $K_1,\dots,K_{2^n}$ be the constituents on $B_1,\dots,B_n$. For any s, $1 \le s \le 2^n$, define $M(K_s) = (p)_1(p)_2\cdots(p)_n$, where $(p)_i = p_i$ if B_i appears unnegated in K_s, and is $q_i(= 1-p_i)$ if it appears negated. It is readily checked that M is a probability model. Taking for example $n = 2$ with $M(K_i)$ as shown,

K_i	$M(K_i)$
$B_1 B_2$	$p_1 p_2$
$B_1 \overline{B}_2$	$p_1 q_2$
$\overline{B}_1 B_2$	$q_1 p_2$
$\overline{B}_1 \overline{B}_2$	$q_1 q_2$

gives

$$p_1 p_2 + p_1 q_2 + q_1 p_2 + q_1 q_2 = p_1(p_2 + q_2) + q_1(p_2 + q_2)$$
$$= 1$$

and

$$P(B_1) = P(B_1 B_2 \vee B_1 \overline{B}_2)$$
$$= M(B_1 B_2) + M(B_1 \overline{B}_2)$$
$$= p_1 p_2 + p_1 q_2 = p_1.$$

That $\mathrm{Ind}(P, B_i)$ holds is also readily seen. For example, if $n = 3$,

$$P(B_1 B_2 B_3) = p_1 p_2 p_3, \qquad P(B_1 B_2) = p_1 p_2$$

so that

$$P(B_1 B_2 B_3) = P(B_1 B_2) P(B_3) = P(B_1) P(B_2) P(B_3).$$

Note the absence of any need to be concerned with the connection of the infinite product $\prod_{i=1}^{\infty}(1 - p_i)$ with the infinite sum $\sum_{i=1}^{\infty} p_i$, nor the separation of cases of convergence and divergence of the sum which Borel seemed to think was necessary.

In his second illustrative problem Borel made use of an extension to the infinite of the principle of total probabilities. To state what corresponds to it for ON probability logic we introduce the concept of *serial (probabilistic) exclusivity of all ϕ instances*, defined as

$$\text{For all } \nu > 0, \quad P(\textstyle\bigvee_{i=0}^{\nu}\phi) = P(\textstyle\bigvee_{i=0}^{\nu-1}\phi) + P(\phi(\nu)), \tag{5}$$

abbreviated to $\mathrm{Excl}(P, \phi(\nu))$. To appreciate the reason for using 'Excl' for the designation note that by P3 (§3.1),

$$P(\phi \vee \psi) = P(\phi) + P(\psi) \text{ if and only if } P(\phi\psi) = 0,$$

the condition $P(\phi\psi) = 0$ being implied by $\vdash \neg(\phi\psi)$ which expresses that an occurrence of both ϕ and ψ is (logically) excluded.

In contrast to serial independence, serial exclusivity does imply pairwise exclusivity. For example, consider three sentences E, F and G for which by exclusivity

$$P(E \vee F \vee G) = P(E \vee F) + P(G)$$
$$= P(E) + P(F) + P(G). \tag{6}$$

Equating the righthand sides of the equations shows that E and F are mutually exclusive, and the first equation that $E \vee F$ and G are. But then if $E \vee F$ and G are mutually exclusive then so are E and G and also F and G. Hence $P(E \vee G) = P(E) + P(G)$ and $P(F \vee G) = P(F) + P(G)$. As this example shows, from (5) and induction we conclude that any $\phi(\nu)$ and $\phi(\mu)$ with $\mu < \nu$ are mutually exclusive.

Our version replacing Borel's principle is

THEOREM 3.43. (Countable additivity) *For any probability function P,* *if Excl(P, $\phi(\nu)$) then*

$$P((\bigvee i)\phi) = \sum_{i=0}^{\infty} P(\phi).$$

PROOF. Assume the hypothesis. Then proceeding as in the proofs of Theorems 3.41, 3.42,

$$P(\phi_0 \vee \phi_1) = P(\phi_0) + P(\phi_1)$$

$$P(\phi_0 \vee \phi_1 \vee \phi_2) = P(\phi_0 \vee \phi_1) + P(\phi_2)$$

$$= P(\phi_0) + P(\phi_1) + P(\phi_2),$$

so that by induction

$$P(\bigvee_{i=0}^{\nu} \phi) = \sum_{i=0}^{\nu} P(\phi).$$

On taking limits ν going to infinity and using P4, we have

$$P((\bigvee i)\phi) = \sum_{i=0}^{\infty} P(\phi).$$

$\square$

In each of the foregoing theorems the result was independent of the particularities of the ON language since no primitive recursive functions were referred to. The following theorem involves an ON language that includes a primary function p with $p(i) = i$ and a subordinate function $q(i, j)$ with $q(i, j) = i + j$. It thus has atomic formulas A_i and A_{i+j}. (See EXAMPLE 2, §2.1.) As i and j range over the numerals ϕ's atomic sentences then are $A_0, A_1, \ldots, A_\nu, \ldots$. The theorem is of interest not only for probability theory but also for our quantifier probability logic in that it involves an instance of a multiply-quantified sentence with both kinds of quantifiers. It also serves, as we shall see, to answer Borel's *Problème III*.

THEOREM 3.44. (Analogue of the Borel-Cantelli lemma) *Let $\mathcal{Q}$ be an* *ON language with primitive recursive functions $p(i) = i$ (primary) and* *$q(i, j) = i + j$ (subordinate) and P a probability function on $\mathcal{Q}$.*
(a) If $\sum_{k=0}^{\infty} P(A_k) < \infty$ then

$$P((\bigwedge i)(\bigvee j)A_{i+j}) = 0.$$

(b) If $\sum_{k=0}^{\infty} P(A_k) = \infty$ and $\mathrm{Ind}(P, A_\nu)$, then

$$P((\bigwedge i)(\bigvee j)A_{i+j}) = 1.$$

PROOF. For (a), by Theorem 3.11

$$P((\bigvee j)A_{\nu+j}) \leq \sum_{j=0}^{\infty} P(A_{\nu+j}) = \sum_{k=\nu}^{\infty} P(A_k).$$

Since, by hypothesis, $\sum_{k=0}^{\infty} P(A_k) < \infty$ then the right-hand side of this inequality has limit 0 as $\nu \to \infty$, so that

$$\lim_{\nu \to \infty} P((\bigvee j)A_{\nu+j}) = 0. \tag{7}$$

Continuing, since

$$\vdash \bigwedge_{i=0}^{\nu}(\bigvee j)A_{i+j} \to (\bigvee j)A_{\nu+j}$$

then (by §3.1 (b))

$$P(\bigwedge_{i=0}^{\nu}(\bigvee j)A_{i+j}) \leq P((\bigvee j)A_{\nu+j}) \tag{8}$$

On taking limits $\nu \to \infty$ and using P4 on the left side produces

$$P((\bigwedge i)(\bigvee j)A_{i+j} \leq \lim_{\nu \to \infty} P((\bigvee j)A_{\nu+j}).$$

Then with (7) this yields

$$P((\bigwedge i)(\bigvee j)A_{i+j}) = 0,$$

so establishing part (a) of the theorem.

For part (b) we have, by P2 and §3.1(c),

$$P((\bigwedge i)(\bigvee j)A_{i+j}) = 1 - P((\bigvee i)(\bigwedge j)\neg A_{i+j}); \tag{9}$$

and by Theorem 3.11

$$P((\bigvee i)(\bigwedge j)\neg A_{i+j}) \leq \sum_{i=0}^{\infty} P((\bigwedge j)\neg A_{i+j}). \tag{10}$$

Since by hypothesis $\mathrm{Ind}(P,\ A_\nu)$, so also $\mathrm{Ind}(P,\ \neg A_\nu)$ (Theorem 3.42). Then by Theorem 3.41

$$P((\textstyle\bigwedge j)\neg A_{i+j}) = \prod_{j=0}^{\infty} P(\neg A_{i+j}) = \prod_{k=i}^{\infty} P(\neg A_k)$$

$$= \prod_{k=i}^{\infty}(1 - P(A_k)). \tag{11}$$

By well-known theorems if $\sum_{k=i}^{\infty} P(A_k) = \infty$ then the infinite product in (11) has the value 0.[4] Since our hypothesis is that $\sum_{k=0}^{\infty} P(A_k) = \infty$, it is also the case for any i that $\sum_{k=i}^{\infty} P(A_k) = \infty$. Hence (11) implies that for any j, $P((\bigwedge i)\neg A_{i+j}) = 0$. Then by (10) and (9) the result follows. □

The above is the theorem that answers the interesting question Borel asks in his *Problème III*: What is the probability that success appears infinitely often, it being assumed that the chance of success at trial n is p_n [and the trials are independent]? If A_n represents the event of success at the nth trial then $(\bigvee j)A_{i+j}$ states that success occurs at the ith or later trial. Accordingly $(\bigwedge i)(\bigvee j)A_{i+j}$ is the statement that at every trial success occurs at that or a later one—hence that there can't be only a finite number of successes. Borel's result (i.e., Theorem 3.44) is that there is an "all or nothing" solution, the probability is 0 or 1, 0 if the p_n's sum to a finite value, 1 if they sum to ∞.

Our next theorem is an ON version of one, due to Fréchet (*1935*, p. 386), extending to the infinite case a result of Boole's.

THEOREM 3.45. (Boole-Fréchet) *Let P be a probability function on an ON language $\mathcal{Q}$ and $(\bigwedge i)\phi$ a formula in $\mathcal{Q}^{cl}$. Let $\min_i\{P(\phi(i))\}$ abbreviate $\lim_{\nu\to\infty}[\min\{P(\phi(0)),\ldots,P(\phi(\nu))\}]$, then*

$$\max\{0,\ 1 - \sum_{i=0}^{\infty} P(\neg\phi)\} \le P((\textstyle\bigwedge i)\phi) \le \min_i\{P(\phi(i))\}.$$

PROOF. For $(\bigwedge i)\phi$ in $\mathcal{Q}^{cl}$,

$$\vdash \textstyle\bigwedge_{i=0}^{\nu}\phi \to \phi(k) \quad \text{for } k = 0,\ldots,\nu.$$

---

[4]Or, one can prove it independently by taking the log of the product and then using the inequality $\log(1 - a) < -a$ for $0 < a < 1$, with $a = P(A_k)$.

Hence by §3.1 (b),

$$P(\bigwedge_{i=0}^{\nu}\phi) \leq \min\{P(\phi(0)), \ldots, P(\phi(\nu))\},$$

and on taking limits,

$$\lim_{\nu \to \infty} P(\bigwedge_{i=0}^{\nu}\phi) \leq \lim_{\nu \to \infty} \min\{P(\phi(0)), \ldots, P(\phi(\nu))\}$$

that is, by P4,

$$P((\bigwedge i)\phi) \leq \min_{i}\{P(\phi(i))\}. \tag{12}$$

Continuing, by §3.1 (g),

$$P((\bigwedge i)\phi) = 1 - P((\bigvee i)\neg\phi),$$

and by Theorem 3.11

$$P((\bigvee i)\neg\phi) \leq \sum_{i=0}^{\infty} P(\neg\phi),$$

so that

$$P((\bigwedge i)\phi) \geq 1 - \sum_{i=0}^{\infty} P(\neg\phi)$$

$$\geq \max[0, \ 1 - \sum_{i=0}^{\infty} P(\neg\phi)],$$

which provides the other inequality. $\square$

§3.5. Kolmogorov probability spaces

This section is devoted to a comparison of the set-theoretical mathematical representation of probability theory, one almost universally employed by probabilists, and that of our logical representation which uses probability functions on ON languages.

The original axiomatic formulation appeared in *Kolmogorov* 1933. We present a slightly modified equivalent one (see, e.g., *Shiryayev* 1995, Chapter II) and, as with Kolmogorov's original version, it is an abstract (set theory) structure with meaningful names attached to the compounds, serving to indicate intended application for probability theory.

The basic notion is that of a *probability space*. This is an ordered triple $< \Omega, \mathcal{A}, P >$, where Ω is a non-empty set (the *sample space of elementary events*), $\mathcal{A}$ is a set of subsets of Ω (the *chance* events) and P (for *probability*) is a set-function on $\mathcal{A}$ into the reals. These are assumed to obey the axioms:

K1. $\mathcal{A}$ is a σ-algebra of subsets of Ω.

K2. P is a probability on elements of $\mathcal{A}$.

Axiom K1 postulates that, in addition to the usual set operations on $\mathcal{A}$, any countable (denumerable) union of sets is also a set of $\mathcal{A}$. As for axiom K2, that P is a *probability* on $\mathcal{A}$ means that $P(\Omega) = 1$ and that P is *countably additive*, i.e., that the P value of a (finite or) denumerably infinite union of pairwise disjoint sets is equal to the sum of the P values of the sets. (See Theorem 3.43 for the comparable property for a probability function on an ON language.)

In §4.7 of SPL we showed that, as far as finite stochastic situations is concerned, results obtained by either sentential probability functions or Kolmogorov probability spaces are intertranslatable, and give equivalent results. However, when the probability space Ω is denumerable differences arise, for then the set of subsets of $\mathcal{A}$ of Ω could be non-denumerable, while an ON language can have only denumerably many sentences. We contrast these two viewpoints by comparing their respective treatments of Bernoulli's law of large numbers. This is an instance of Borel's denumerable probabilities where the number of trials is infinite but the number of outcomes of a trial is finite (two in this case).

A trial here is a chance experiment with two possible outcomes, success or failure, having respective probabilities p and $q\,(=1-p)$, the same at each trial. The trials are independent. What happens to the relative frequency of the number of successes to the number of trials as the trials are indefinitely repeated?

First a preliminary calculation. For n of these Bernoulli trials (as they are called) an outcome is of the form $(B_1,\ldots,B_n)$ where each B_i is either success of failure. Let ω stand for an arbitrary one of the 2^n outcomes and $P^{(n)}(\omega)$ its probability. Then

$$P^{(n)}(\omega) = p^k q^{n-k}, \tag{1}$$

where k is the number of successes. The superscript '(n)' on P is to remind us that our sample space consists of the 2^n n-tuples of B_i's. Straightforward mathematical calculations show (see, e.g., *Renyi 1970*, p. 158, or *Shiryayev 1984*, p. 47 (5)) that for ϵ a positive number the sum of probabilities of the form (1) satisfies the estimate

$$\sum_{\{\omega:|\frac{k}{n}-p|>\epsilon\}} P^{(n)}(\omega) < \frac{pq}{n\epsilon^2} < \frac{1}{4n\epsilon^2} \tag{2}$$

the sum being taken over all ω's satisfying $|\frac{k}{n}-p| > \epsilon$. Since the ω's are mutually exclusive the sum can (by additivity) be taken as the probability of this outcome, i.e., as

$$P^{(n)}(\{\omega : |\frac{k}{n} - p| > \epsilon)\}) < \frac{1}{4n\epsilon^2}. \tag{3}$$

One is tempted to say that this indicates that the chances of the relative frequency k/n differing from the probability p can be made vanishingly small by taking n large enough. But as $P^{(n)}$ changes with n it isn't clear that (3) justifies that conclusion.

We present a Kolmogorov probability space (set-theoretic) treatment of the question. The first step is to define a suitable probability space $< \Omega, \mathcal{A}, P >$. The set Ω is taken to be the set of all possible infinite sequences

$$B_1,\ldots,B_n,\ldots \tag{4}$$

where $B_n\ (n = 1, 2, \ldots)$ is either success or failure. This Ω has $2^{\aleph_0}$ elements. Defining $\mathcal{A}$ and P is done in a two step procedure.

Two sequences of the form (4) are said to be *n-initial equivalent* if their initial segments of length n are identical. This relation (clearly an equivalence relation) divides Ω into equivalence classes (sets). We refer to these equivalence classes as being sets of the n-th *order*. Let these sets be denoted by $L_{b_1 b_2 \cdots b_n}$ where b_i is 1 if the i-th entry in the initial segment is *success* and 0 if it is *failure*. For any n, each of the 2^n possible sets of the n-th order are mutually exclusive, their union is Ω, and $L_{b_1 b_2 \cdots b_n} = L_{b_1 b_2 \cdots b_n 1} \cup L_{b_1 b_2 \cdots b_n 0}$. Then a set $\mathcal{A}_0$ is defined as the closure of these sets (all the equivalence classes for all n) under finite unions, intersections, and complements with respect to Ω. Next a function P_0 is obtained by assigning to each equivalence class the value $p^k q^l$, where k is the number of occurrences of *success* in the initial segment and $l (= n - k)$ the number of *failure*, and (b) extending these values to all of $\mathcal{A}_0$. This extension is accomplished by expressing any element of $\mathcal{A}_0$ (excepting the empty set) as a union of equivalence classes of the same order (by the same technique used to obtain a Boolean normal form) and taking P_0 as additive over such unions.

It is straightforward to show that $\langle \Omega, \mathcal{A}_0, P_0 \rangle$ is a finitely additive probability space, i.e., satisfies Kolmogorov's axioms with axiom K2 reduced to finite additivity. However $\langle \Omega, \mathcal{A}_0, P_0 \rangle$ is not yet adequate since $\mathcal{A}_0$ is not a σ-algebra. From the set-measure-theoretic viewpoint a σ-algebra is needed to insure "complete freedom of operating with events without the fear of arriving at results which possess no probability." (*Kolmogorov* 1933, 15). Such freedom can be had: there are mathematical theorems that guarantee the existence of a least σ-algebra $\mathcal{A}$ containing $\mathcal{A}_0$ and that P_0 can be uniquely extended to a function P on $\mathcal{A}$ which is σ-additive (*ibid.*, 15–16). Using these theorems produces the desired $\langle \Omega, \mathcal{A}, P \rangle$. Although, Kolmogorov notes, one looks at the elements of the σ-algebra $\mathcal{A}$ as "ideal events" which may not correspond to anything in experience, nevertheless, it is argued, when a deduction, carried out by using such idealized constructs leads to the determination of the probability of a real event (one in $\mathcal{A}_0$), then this determination would, evidently, be automatically unobjectionable from an empirical standpoint.[5]

---

[5]Mengen aus $\mathfrak{BF}$ betrachten wir also im allgemeinen nur als „ideelle Ereignisse", welchen in der Erfahrungswelt nichts entspricht. Wenn aber eine Deduktion, die die Wahrscheinlichkeiten solcher ideelen Ereignisse gebraucht, zur Bestimmung der Wahr-

To compare this set-theoretic treatment of the law of large numbers with a logic-theoretic one requires replacing the probability space $< \Omega, \mathcal{A}, P >$ with an appropriate $< \mathcal{Q}, M >$, where $\mathcal{Q}$ is an ON language and M a probability model for $\mathcal{Q}$. For $\mathcal{Q}$ we choose one whose atomic sentences are $A_1, \ldots, A_n, \ldots$, where A_n $(n = 1, 2, \ldots)$ expresses (via the meaning we are now choosing) that the result of trial n is *success*, and $\neg A_n$ that the result is *failure*. Replacing Ω will be the constituents on $A_1, \ldots, A_n$, designated $K_{b_1 b_2 \cdots b_n}$, $(n = 1, 2, \ldots)$ where b_i is 1 if A_i appears in $K_{b_1 b_2 \cdots b_n}$ unnegated and b_i is 0 if it appears negated. The probability model (§3.1, definition following Theorem 3.11) chosen, M, is defined by setting $M(K_{b_1 b_2 \cdots b_n}) = p^k q^{n-k}$, k being the number of unnegated A_i in $K_{b_1 b_2 \cdots b_n}$ and $n - k$ the number of negated A_i. [See NOTE, next paragraph] By Theorem 3.21 M determines a unique probability function P on $\mathcal{Q}^{cl}$. Since P is defined for every sentence of $\mathcal{Q}$ we have (1) holding for this unique P no matter which sequence of trials is considered and no matter which n. The argument that produced (3) then holds with $P^{(n)}$ replaced with P which doesn't change with n.

NOTE. To see that this assignment M is a probability model according to the definition in §1.2 II, consider the simple case of $n = 3$:

$$M(A_1 A_2 A_3) +$$
$$M(A_1 A_2 \overline{A}_3) + M(A_1 \overline{A}_2 A_3) + M(\overline{A}_1 A_2 A_3) +$$
$$M(\overline{A}_1 \overline{A}_2 A_3) + M(A_1 \overline{A}_2 \overline{A}_3) + M(\overline{A}_1 A_2 \overline{A}_3) +$$
$$M(\overline{A}_1 \overline{A}_2 \overline{A}_3)$$
$$= p^3 + 3p^2 q + 3pq^2 + q^3$$
$$= (p + q)^3 = 1,$$

where the last equality comes from $p + q = 1$. The other probability model properties are readily verified.

How does the logic-theoretic approach to probability theory, treating stochastic problems involving infinite trials and/or outcomes, do without an assumption such as σ-additivity? It doesn't need one—the property is built into the semantics. Note that the definition of a probability function

---

scheinlichkeit einer reellen Ereignisses aus $\mathfrak{F}$ führt, wird diese Bestimmung offensichtlich automatisch auch vom empirischen Standpunkte aus einwandfrei sein. (*ibid.* p. 16).

assumes that there is a value for each closed formula of an ON language (§3.1, P1–P4). The Main Theorem (§3.2) shows that for any probability model there is a unique function that extends M so as to be a probability function P_M on the language. For such functions " σ-additivity" is provable (Theorem 3.43).

§3.6. Logical consequence in probability logic

We open this section with a brief historical remark on the two aspects of formal logic—linguistic structure and associated semantics— and then proceed to a summary discussion of our views on logical consequence.

An early, and for a long time unknown, formalization of logical principles in terms of linguistic structure occured in an unpublished manuscript of Leibniz of 1686. In it there is an axiomatization of Aristotelian-medieval-scholastic term logic. (*Generales Inquisitiones de Analysi Notionum et Veritatum*, accessible in *Couturat* 1903, pp. 356–399.) Much later, not until *Frege* 1879, do we have axiomatizations of contemporary sentential logic and also quantifier logic for predicate languages.

As for semantic ideas they do occur inchoately throughout the history of logic but not until the 1920's and 1930 were they singled out, systematized and used as the basis for justifying validity of syntactically formulated principles. For quantifier logic this semantics is considerably simplified by employing the ON form. And, rather than *validity of a formula* as the key notion, we are using the slightly more general *(valid) logical consequence of a formula from a finite set of formulas.*

As we have seen in §1.1 this definition for sentential verity logic is simple: a formula ψ is a logical consequence of formulas $\phi_1, \ldots, \phi_m$ if, for every assignment of verity values to the atomic components, i.e., for every verity model, when $\phi_1, \ldots, \phi_m$ have the value 1 so does ψ. This requires knowing how such values are transmitted to a formula from assignments to its atomic components. This transmission is accomplished, in §1.1, by means of verity functions. Commonly this is the task of connective-defining truth tables.

But tables, being finite, do not generalize as conveniently as functions. The definition of a model in §1.1 refers to an assignment of verity values to all atomic sentences of $\mathcal{S}$, even though any (sentential) inference scheme involves only a finite number of them. Although not essential for sentential logic this simultaneous assignment to all atomic sentences is needed for ON quantifier logic (§2.1).

When we come to probability logic, although the formal language is the same as that for verity logic, there is a marked change in the semantics. First of all verity functions give way to probability functions. These functions, whose domain is the same as that for verity functions, namely the sentences of $\mathcal{S}$, have a range that is not the $\{0, 1\}$ of verity functions but the unit interval $[0, 1]$. Algebraic(-arithmetic) operations on probability function values include not only max and min but also addition, subtraction and multiplication. The algebraic relations '$\leq$' and '$\geq$' are also used. For sentential probability logic the basic semantic sentence forms are '$P(\phi) \in \alpha$', ϕ a closed formula and α a subset of $[0, 1]$. While for verity logic the simple algebra over $\{0, 1\}$ (instead of linguistic combinations of *false* and *true*) was just a convenience in expressing semantic concepts, for probability logic having the algebra over $[0, 1]$ is essential for aleatoric usage of the logic. Probability functions govern the transmission of probability values through the formal language but not in the simple manner that verity functions do. Contrast, for example,

$$V(\phi \vee \psi) = \max\{V(\phi), V(\psi)\} \quad \text{with}$$
$$P(\phi \vee \psi) = P(\phi) + P(\psi) - P(\phi\psi),$$

with the values of V in $\{0, 1\}$ and those of P in $[0, 1]$.

In the generalization of verity to probability logic not only is there a change in semantics from verity functions to probability functions but also the notion of a model is extended. For a probability model assignments (of values from $[0, 1]$) are made not to atomic sentences but to constituents constructed on atomic sentences. And, just as in verity logic, if value 1 is assigned to an atomic sentence A, then 0 has to be assigned to $\neg A$, so values assigned to constituents have to be done appropriately (as described in §1.2 II). Once probability values are assigned to all constituents by a model then every sentence has a determined value, namely that which the probability function determines from its complete disjunctive normal form. That this notion of a probability model does indeed generalize that of model for verity logic is seen by noting that a probability model in which 1 is assigned to a constituent on n atomic sentences, and 0 to all others on these n sentences, is equivalent to a verity model assignment on these n sentences (SPL, Theorem 0.12).

We consider logical consequence in some more detail. Customarily the formal language—its linguistic structure—is precisely specified, but not so the semantic language which is considered to be *meaningful*. In the case of sentential verity logic we have seen that the several ingredients, namely, '$V(\phi) = 1$', the max and min of pairs of V values, semantically understood simple connectives ('and', 'if, then') and quantification over verity models serve to characterize logical consequence. (For sentential logic only a finite number, namely 2^n, of verity models if there are only n atomic sentences involved, need to be examined to determine if there is a valid logic consequence.)

When we come to sentential probability logic probability functions are introduced and verity logic's '$V(\phi) = v$' gives way to '$P(\phi) \in \alpha$'. There are also the additional algebraic operations of addition, subtraction and multiplication as well as the relations '$\leq$' and '$\geq$'. While for this sentential probability logic there isn't a simple "truth table" method of determining valid logical consequence there is, nevertheless, an effective procedure for determining the optimal interval for the consequent if the subsets of $[0, 1]$ for the antecedents are explicitly given subintervals with rational end points (SPL, §4.6).

Adjoining quantifiers to the syntax of sentential language so as to produce (the ON form) of verity quantifier language brings with it an enlargement of the semantic language, namely an extending of the notion of a verity function with the inclusion of the operations min and max of an infinite sequence of verity values (§2.1). A model here is still the same as for sentential language—an assignment of verity values to the atomic sentences—except that for ON language the atomic sentence symbols have specified numeral subscript constants. Any model so-defined determines a unique verity function on the closed sentences of an ON language (Theorem 2.11).

In extending probability logic to the quantifier level the definition of a probability function is augmented with an additional property, P4 of §3.1: the probability of a general $\bigwedge$-quantification is the limit of the probability of the corresponding finitization as the number of conjunctive components goes to infinity. Similarly for the $\bigvee$-quantification, *mutatis mutandis*. Thus limiting processes in the real number system come into the semantics. A probability model for on ON language $\mathcal{Q}$ is the same for a sentential language $\mathcal{S}$ with ON's more general atomic sentences replacing those of $\mathcal{S}$. The probability values assigned by the model (to constituents) are uniquely extendable so as to be a probability function on the closed formulas of $\mathcal{Q}$ (Theorem 3.21.) Thus the semantics for quantifier probability logic includes notions such as $P((\bigvee i)\phi)$, $P((\bigwedge i)\phi)$ and more complicated constructs such as $\mathrm{Ind}(P, \phi(\nu))$ and $\mathrm{Excl}(P, \phi(\nu))$. There are also items involving limits, e.g., $\prod_{i=0}^{\infty} P(\phi)$, $\sum_{i=0}^{\infty} P(\phi)$ and $\min_i\{P(\phi)\}$.

All this probability-logic development just summarized has been based on a syntax language unchanged from that for verity logic. For conditional-probability logic we extended the formal syntax language with the introduction of the suppositional which when introduced into sentential verity logic, the resulting semantics was of a 2-to-3 valued nature. Our next chapter, Chapter 4, considers its introduction into quantifier logic.

§3.7. Borel's denumerable probabilities defended

In *Barone & Novikoff* 1978—a history of the axiomatic formulation of probability from Borel to Kolmogorov—we find praise for Borel's "landmark paper initiating the modern theory of probability" (i.e., *Borel* 1909), and yet severe criticism for the "many flaws and inadequacies in BOREL's reasoning".

We devote this section to showing that from the ON probability logic perspective much of the inadequacies in Borel's conception of probability theory can be filled in. Also, one can provide support for his philosophical position regarding the infinite in mathematics, a position to which he hoped his paper was a contribution.

As the following excerpt shows, Barone & Novikoff couldn't conceive it otherwise but that probability theory had to be based on measure theory (*1978*, p. 125):

> Any adequate discussion of the contents of BOREL's landmark paper (which we shall refer to as BOREL (1909)) is of necessity delicate and somewhat detailed. The reason for this is the ironical circumstance that BOREL, the unquestioned founder of measure theory, attempted in 1909 to found a new theory of "denumerable probability" *without* relying on measure theory. The irony is further compounded in the light of BOREL's paper of 1905 which identified "continuous probability" in the unit interval with measure theory there. Consequently, we are at great pains both to establish and comprehend BOREL's reluctance in 1909 to accept the underlying role of countable additivity in his new theory.

Instead of countable additivity Borel relied, they contended, on a new principle which they called "countable independence":

> ... We propose to use the phrase "*countable independence*" for the principle that BOREL explicitly introduced and on which all of his results are based. This is the assertion, usually taken as a hypothesis, that a given collection of events $B_1, B_2, \ldots, B_n, \ldots$ satisfy

$$P(\bigcap_1^\infty B_i) = \prod_1^\infty P(B_i). \tag{2.1}$$

> When the collection of events is finite, the corresponding principle was known as the "loi des probabilités composées", although the notation for set intersection was not generally employed. BOREL assumed the principle if the events B_i referred to different trials for different i and, most important, assumed it to hold even if the index were infinite (*1978*, p. 130).

Borel didn't use a symbol for probability (e.g., 'P') or specify in detail its properties. Moreover, he didn't conceive of the "events B_i" (also unsymbolized by Borel) as being, or being represented by, sets so that he would

not have used the set-intersection symbol '$\bigcap$'. It is then feasible to enter-
tain an interpretation, or clarification, of his "new principle" (as Barone &
Novikoff refer to it) alternative to their (2.1).

Their opinion is that Borel came to this principle in analogy with his generalization of the idea of length on the real line, referring to it as "BOREL's
earlier, profoundly important discovery of the theory of measure". This
theory extends the idea of length from that of the sum of a finite number
of disjoint (finite) intervals to that of a denumerable sum via ('l' standing
for 'length of' and '$\bigoplus$' for the union of disjoint sets)

$$l(\bigoplus_{i=1}^{\infty} B_i) = \sum_{i=1}^{\infty} l(B_i). \tag{1}$$

Since $\bigoplus_{i=1}^{\infty} B_i$ need not be an interval this clearly involves an extension of
the idea of length. It is not clear why Barone & Novikoff think that their
interpretation of Borel's "fundamental new principle", i.e., (2.1), was suggested to Borel ("must have acted powerfully on BOREL") rather than (1)
suggesting the more closely related extension of total probability (countable
additivity),

$$P(\bigoplus_{i=1}^{\infty} B_i) = \sum_{i=1}^{\infty} P(B_i), \tag{2}$$

for mutually disjoint sets (exclusive events).

It seems to us, however, that Borel happened to use this extension of
the law of compound probability in his first illustrative example and hence
would have discussed it in some detail, but he also referred to the extension
of the law of total probability (i.e., countable additivity) in his solution
of *Problème II* and, by way of justification, casually saying "A reasoning
analogous to that which we used in *Problème I* allows the application ... of
the [extension of the] principle of total probabilities" (*Borel* 1909, p. 250).

In their discussion of Borel's calculation of A_0 as yielding $(1 - p_1)(1 -
p_2) \cdots (1 - p_n) \cdots$ (see our §3.3 above) they state (p. 134):

In fact, what BOREL is skimming over is the limit relation

$$\lim_{n \to \infty} P(\bigcap_{1}^{n} B_i) = P(\bigcap_{1}^{\infty} B_i). \tag{4.2}$$

An assumed independence assures additionally that if each B_i has probability $q_i \, [= 1 - p_i]$, then

$$P(\bigcap_1^n B_i) = \prod_1^n P(B_i) = \prod_1^n (1 - p_i).$$

The limit relation (4.2), however, has nothing to do with independence; it is one of the many consequences, and even equivalent forms, of countable additivity. The limit relation (4.2) is, for BOREL, both too desirable to be false and too evident to require discussion or elaboration. Indeed, since he nowhere employs a notation for the algebra of sets or even for sets themselves or for set functions, it would not have been easy for him to state explicitly. Had he employed the symbolism $P(\bigcap_1^\infty B_i)$, perhaps he might have been driven to question the domain of the set function $P(\cdot)$ just as he had earlier questioned and extended the domain of "length" for point sets in $[0,1]$.

In response to this criticism our probability logic solution in §3.4 fills in this "skimming over" by having P specified via an appropriate probability model (Theorem 3.21). Then P, being a probability function (on sentences, not sets), satisfies P4 (§3.1) and indeed

$$\lim_{n \to \infty} P(\bigwedge_{i=1}^n B_i) = P((\bigwedge i)B_i).$$

This doesn't require independence of the B_i. What does require (serial) independence is, in virtue of Theorem 3.41, the property

$$P((\bigwedge i)B_i) = \prod_{i=0}^\infty P(B_i).$$

Another of their criticisms concerns Borel's concept of "denumerable probabilities". They preface their criticism with a long quotation from the opening remarks of Borel's *1909* of which the following is its final portion.

    ... The cardinality of denumerable sets alone being what we may know in a positive manner, the latter alone intervenes *effectively* in our reasonings. It is clear, indeed, that the set of analytic elements that can be actually defined and considered can be only a denumerable set; I believe that this point of view will prevail more and more

every day among mathematicians and that the continuum will prove
to have been a transitory instrument, whose present-day utility is not
negligible (we shall supply examples at once), but it will come to be
regarded only as a means of studying denumerable sets, which con-
stitute the sole reality that we are capable of attaining (*Borel* 1909,
247–248).

Following this Barone & Novikoff remark:

> These opening words indicate that BOREL believes that the set of
> possible outcomes which he will discuss and which in modern terms is
> his sample space, is denumerable. Nothing could be more misleading:
> the sample spaces he discusses are always denumerably infinite prod-
> ucts of finite, or at most denumerably infinite, factor spaces. Indeed,
> even the simplest of these, the denumerable CARTESIAN product of
> 2-point spaces, is nondenumerable, as had been shown earlier by G.
> CANTOR. (*Barone & Novikoff* 1978, 131–132.)

The argument here is cogent only if one agrees with their tacit assumption that only by use of a Kolmogorov probability space (§3.5) can one
satisfactorily treat probability theory with infinitly many occurrences of
events.

Our discussion of the example of the Law of Large Numbers in §3.5 implies that ON probability logic offers a non-nondenumerable alternative.
Instead of $\langle \Omega, \mathcal{A}, P \rangle$, both Ω and $\mathcal{A}$ being nondenumerable and P a probability function on $\mathcal{A}$, ON probability logic employs $\langle \mathcal{Q}, M \rangle$. Here $\mathcal{Q}$ is
a countable (quantifier) language and M is a probability model on $\mathcal{Q}$. It
assigns values from $[0, 1]$ to the constituents $K_{b_1 \cdots b_n}$ $(n = 1, 2, \ldots)$ constructed on the atomic sentences of $\mathcal{Q}$. Everything here is denumerable
except for $[0, 1]$. And if the assignments that M makes to the sentences
$K_{b_1 \cdots b_n}$ (namely $p^k q^{n-k}$) are so chosen as to avoid indefinable reals—as
would be the case if p were required to be such a real—then there is con-
cordance with Borel's ideas.

CHAPTER 4

CONDITIONAL-PROBABILITY AND QUANTIFIERS

In the preceding chapters, starting with verity sentential logic, we engaged in a successive widening of the scope of logic. The first step was to a probability logic whose formal language was the same as that of verity sentential logic but with the latter's two semantic values, 0 and 1, increased to encompass the real numbers from 0 to 1. Then, in preparation for the introduction of conditional probability, a new connective, the suppositional, was introduced into sentential language making use of a new semantic value u ("undetermined"). Not only was the suppositional of use in defining conditional probability, but apart from this it possessed interesting properties as a logical notion in its own right. In §2.4 suppositional logic was enlarged to include quantifiers or, equivalently, quantificational logic was enlarged with the suppositional. Chapter 3 was devoted to (unconditional) probability on quantifier languages.

Finally, in this chapter we conclude the succession of enlargements so as to have conditional-probability for (somewhat limited) quantifier languages. The concept of the suppositional in probability logic with quantifiers is so new that we have but one application, a fresh approach to a resolution of the paradox of confirmation.

§4.1. Conditional-probability in quantifier logic

The conditional-probability function P^* was introduced in §1.5 as an extension of the sentential probability function P. It coincided with P on formulas of S (elements of S^u not containing an occurrence of '⊣'). In general, for an element of S^u it supplied a value in $[0,1]$ or, failing that, the indefinite (numerical) constant c as a value. Although we call

P^* a conditional-probability function, it is not a probability function as defined in §1.2. In §§3.1, 3.2 it was shown how a meaningful notion of (unconditional) probability can be provided for ON (quantifier) languages. Since quantifications are for us, intuitively, "infinitely" long conjunctions or disjunctions, it was natural to define the probability of such formulas as the limit of the probability of finitely many terms of the conjunctions, or disjunctions, as the number goes to infinity, as seen by virtue of P4 (§3.1) extended to all prenex formulas by the MAIN THEOREM (Theorem 3.21). This comports with the early historical ideas of Borel described in §3.3. We continued to use the same unadorned symbol 'P', with the domain being extended from $\mathcal{S}$ to $\mathcal{Q}^{cl}$, and subject to the additional property P4.

Here we now further enlarge our probability notion with the inclusion of both the suppositional and quantifiers, the symbol for this enlarged notion being '$P^\sim$'. The relationships of all these semantic probability functions is depicted in Figure 3, the arrow meaning 'inclusion by'.

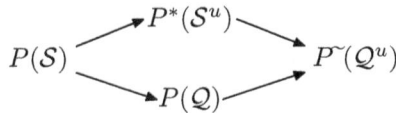

$$P(\mathcal{S}) \quad P^*(\mathcal{S}^u) \quad P^\sim(\mathcal{Q}^u) \quad P(\mathcal{Q})$$

FIGURE 3.

Here '$P(\mathcal{S})$' indicates that the domain of P is $\mathcal{S}$, '$P^*(\mathcal{S}^u)$' that of P^* is $\mathcal{S}^u$, etc. By the conventions adopted in the respectively indicated sections alongside of the inclusions here indicated, we have:

$$\mathcal{S} \subset \mathcal{S}^u \subset \mathcal{Q}^u \quad \text{(by §1.4, by §2.4)}$$
$$\mathcal{S} \subset \mathcal{Q} \subset \mathcal{Q}^u \quad \text{(by §2.1, by §2.4).}$$

In view of the limited application we shall here be making of $P^\sim$ it will suffice to define it just for the kind of formulas of $\mathcal{Q}^u$ involved in the application described in the next section.

DEFINITION OF $P^\sim$

Let P^* be a conditional-probability function as defined in §1.5.

   (i) If Φ is a closed formula of $\mathcal{S}^u$, set

$$P^\sim(\Phi) = P^*(\Phi).$$

(ii) For any closed, quantifier-free (except for the $(\bigwedge i)$ as shown) formulas $(\bigwedge i)\Phi$, Ψ in $\mathcal{Q}^u$, set

$$P^{\widetilde{}}((\textstyle\bigwedge i)\Phi \dashv \Psi) = \begin{cases} \lim_{n\to\infty} P^*((\bigwedge_{i=0}^{n}\Phi) \dashv \Psi) & \text{if the limit exists} \\ c, & \text{otherwise,} \end{cases}$$

where c is (as in §1.5) an unspecified value in $[0, 1]$.

It follows from the properties of P^* (see the remark immediately following the definition of P^* in §1.5) that the value of $P^{\widetilde{}}$ is unchanged if an $\mathcal{S}^u$ formula part in its argument is replaced by a u-equivalent formula.

For Θ in $\mathcal{S}^u$, $\Theta \dashv \mathbf{1}$ and Θ are u-equivalent (EXAMPLE 3, §2.4). Hence they have the same suppositional normal form so that by the definition of P^* (§1.5).

$$P^*(\Theta \dashv \mathbf{1}) = P^*(\Theta).$$

It will then be convenient to define

$$P^{\widetilde{}}((\textstyle\bigwedge i)\Phi) \quad \text{as} \quad P^{\widetilde{}}((\textstyle\bigwedge i)\Phi \dashv \mathbf{1})$$

so that

$$\begin{aligned} P^{\widetilde{}}((\textstyle\bigwedge i)\Phi) &= \lim_{n\to\infty} P^*(\textstyle\bigwedge_{i=0}^{n}\Phi \dashv \mathbf{1}) \\ &= \lim_{n\to\infty} P^*(\textstyle\bigwedge_{i=0}^{n}\Phi). \end{aligned}$$

This extended notion of conditional probability, i.e. $P^{\widetilde{}}$, is applied in our next section to the solution of a long standing and much discussed paradox.

§4.2. The paradox of confirmation

The confirmation paradox arose in the context of philosophical discussion of scientific explanation. The paradox, to use the stock example, is: although 'a is a raven and a is black' seems to confirm (add probability to) 'All ravens are black', yet 'a is not black and is not a raven' seems not to, even though the contraposed conditional 'All non-black objects are non-ravens' is logically equivalent to 'All ravens are black'. Over subsequent decades much has been written, and by many authors, on this paradox.[1] We here propose a resolution based on our quantifier probability logic which includes the suppositional.

The paradox was introduced by Carl Hempel (*1945*, reprinted *1965*) in "a study of the non-quantitative concept of confirmation" illustrating shortcomings of Jean Nicod's concept of confirmation. In his paper he quotes from *Nicod* 1930, p. 219:

> Consider the formula or law: *A entails B*. how can a particular proposition, or more briefly, a fact, affect its probability? If this fact consists of the presence of B in a case of A, it is favorable to the law '*A entails B*'; on the contrary, if it consists of the absence of B in case A, it is unfavorable to this law. It is conceivable that we have here the only two direct modes in which a fact can influence the probability of a law. . . Thus, the entire influence of particular truths or facts on the probability of universal propositions or laws would operate by means of these two elementary relations which we shall call *confirmation* and *invalidation*.

Note that Nicod is concerned with how the probability of a universal proposition of the form 'A entails B' is affected by the truth of a particular proposition [expressing] that the presence of [a] B in the case of [an] A is "favorable", while its absence is "unfavorable". In his discussion Hempel renders Nicod's 'A entails B' as a universal (truth-functional) conditional

$$(x)[P(x) \supset Q(x)].$$

According to this interpretation of Nicod's criterion such a "hypothesis" is confirmed by an object *a* which is a P and a Q, and is disconfirmed by one that is a P but not a Q. Hempel then concludes with: "and (we add this to Nicod's statement) it is neutral or irrelevant, with respect to the hypothesis if it does not satisfy the antecedent." However this addition of

---

[1]See, for example, *Swinburne* 1973 and the more than two dozen references cited in it relating to this topic.

Hempel's to Nicod's statement together with Hempel's interpreting 'entails' to be the truth-functional conditional leads to trouble. Adoption of both produces the paradox, inasmuch as an object that is not a P, whether a Q or not, confirms $(x)[\sim Q(x) \supset \sim P(x)]$, which is logically equivalent to $(x)[P(x) \supset Q(x)]$ and hence not "neutral, or irrelevant".

We propose to avoid this undesirable result by interpreting Nicod's "entails" not by having the truth-functional conditional in $(x)(P(x) \supset Q(x))$ but by the suppositional, i.e., by $(\bigwedge i)(Q(i) \dashv P(i))$. It seems fairly clear that the truth-functional conditional is not as informative as the suppositional. For in the case of truth-functional conditional if its antecedent is false then, independently of the value of the consequent, the conditional is taken as being true. On the other hand with the suppositional if its antecedent is false then the semantic value of the suppositional is u, i.e., undetermined as to truth value. Although this supplies no positive information, at least there isn't the possibility of false information as would be the case with the conditional, which is then considered as being true independently of whether the consequent is true or false. The difference is even more pronounced when dealing with probability semantics.

Before continuing we point out the neither Nicod nor Hempel gave any indication that the notion "probability of a universal sentence" was in need of clarification. Perhaps the earliest recognition of this need (in terms of infinitely many trials of an experiment) was in *Fréchet* 1935. (See his section III, p. 383.) Apparently a formal treatment of the probability of quantifications was first accomplished in *Gaifman* 1964.

Generally in studies on confirmation it is assumed that there is some background information accompanying the evidence (antecedent for a suppositional). Comparable to this we assume that in a given circumstance a probability function is specified appropriately with properties suitable for the circumstances. So assume then that we have a probability function $P^{\smallfrown}[= P^{\smallfrown}_M]$ defined by a model M for an ON language (with suppositional) which includes atomic sentences $B_1, \ldots, B_n, \ldots$ and $R_1, \ldots, R_n \ldots$ with B_i saying "Object a_i is black" and R_i that "Object a_i is a raven." We consider the generalization $(\bigwedge i)(B_i \dashv R_i)$ whose meaning is "Everything that is a raven is black" with the special sense of $\dashv$ which attributes in suppositional semantics the value u (undetermined or unknown as to true or false) to $B_i \dashv R_i$ if R_i is false. What happens, now turning to probability semantics, to the probability of this generalization when the additional information, $B_1 R_1$ is given? Or, in other words, how are $P^\sim((\bigwedge i)(B_i \dashv R_i))$ and $P^\sim((\bigwedge i)(B_i \dashv R_i) \dashv B_1 R_1)$ related?[2] The following theorem shows that if there is any change with this addition it is to increase the probability. The hypothesis of this theorem includes the requirement that neither $P(B_1 R_1)$ nor, for any i, $P(R_i)$ be 0, i.e., that they are not "*a priori*" impossible and that $\lim_{n \to \infty} P^*(\bigwedge_{i=1}^{n}(B_i \dashv R_i))$ exists. These requirements preclude having the outcome of this relation being uninformative since then both $P^\sim$ values, i.e. those before and after adjoining $B_1 R_1$ as additional information, are undefined values in [0,1].

THEOREM 4.21. *Let $P^\sim[= P^\sim_M]$ be a conditional-probability function on Q^u whose atomic sentences include B_i, R_i $(i = 1, \ldots, n, \ldots)$ and such that $P^\sim(B_1 R_1) = P^*(B_1 R_1) = P(B_1 R_1) \neq 0$ and also that $P(R_i) \neq 0$ for any i. Then, assuming that $\lim_{n \to \infty} P^*(\bigwedge_{i=1}^{n}(B_i \dashv R_i))$ exists,*

$$P^\sim[(\bigwedge i)(B_i \dashv R_i) \dashv B_1 R_1] \geq P^\sim[(\bigwedge i)(B_i \dashv R_i)].$$

PROOF. For $P^*[= P^*_M]$, in terms of which $P^\sim$ is defined, we shall establish that

$$P^*[\bigwedge_{i=1}^{n}(B_i \dashv R_i) \dashv B_1 R_1] \geq P^*[\bigwedge_{i=1}^{n}(B_i \dashv R_i)], \qquad (1)$$

from which the result follows by letting $n \to \infty$.

Case 1. $n = 1$.

We have by (11) in §1.5

$$P^*[(B_1 \dashv R_1) \dashv B_1 R_1] = P^*(B_1 \dashv B_1 R_1)$$

and by (ii) in §1.5

$$= \frac{P(B_1 R_1)}{P(B_1 R_1)}$$

---

[2]Since in the ensuing discussion we shall be considering objects (e.g., ravens, black things, etc.), we shall assume that our quantifiers range over numerals starting with 1 rather than 0, as it seems peculiar to have the first one in a list of objects being referred to as the "zeroth" one.

which is

$$\geq P^*(B_1 \dashv R_1).$$

Case 2. $n > 1$.

By the same argument used in §2.4 to establish EXAMPLE 2, there are formulas C and A in $\mathcal{S}$ such that

$$\textstyle\bigwedge_{i=1}^{n-1}(B_{i+1} \dashv R_{i+1}) \equiv_u C \dashv A. \tag{2}$$

Then with the use of (2) applied to the right-hand side of

$$P^*[\textstyle\bigwedge_{i=1}^{n}(B_i \dashv R_i)] = P^*[\textstyle\bigwedge_{i=1}^{n-1}(B_{i+1} \dashv R_{i+1})(B_1 \dashv R_1)]$$

we obtain

$$= P^*[(C \dashv A)(B_1 \dashv R_1)]$$

so (that by (9) in §1.5)

$$= P^*[CB_1 \dashv (AR_1 \vee A\overline{C} \vee R_1\overline{B}_1)]$$

and hence by §1.5 (ii),

$$= \frac{P(CB_1AR_1)}{P(AR_1 \vee A\overline{C} \vee R_1\overline{B}_1)}. \tag{3}$$

Also

$$
\begin{aligned}
P^*[\textstyle\bigwedge_{i=1}^{n}(B_i \dashv R_i) \dashv B_1R_1] &= P^*[\textstyle\bigwedge_{i=1}^{n-1}(B_{i+1} \dashv R_{i+1})(B_1 \dashv R_1) \dashv B_1R_1] \\
&= P^*[(C \dashv A)(B_1 \dashv R_1) \dashv B_1R_1] && \text{by(2)} \\
&= P^*[(C \dashv A) \dashv B_1R_1] && \text{Thm. 1.51} \\
&= P^*(C \dashv AB_1R_1) && \text{§1.5 (11)} \\
&= \frac{P(CAB_1R_1)}{P(AB_1R_1)}. && \text{(4)}
\end{aligned}
$$

Since in (3) and (4) the numerators on the right are the same but for the denominators that of (3) is equal to or larger than that of (4), we have

$$P^*[\textstyle\bigwedge_{i=1}^{n}(B_i \dashv R_i) \dashv B_1R_1] \geq P^*[\textstyle\bigwedge_{i=1}^{n}(B_i \dashv R_i)],$$

establishing (1) from which, by taking limits, the conclusion follows. □

Thus the $(P^\frown)$ conditional-probability that everything that is a raven is black, on the supposition that object a_1 is a black raven, can't be less than the $(P^\frown)$ probability that everything that is a raven is black.

But suppose the additional information is that there is a non-black raven. Then we have

THEOREM 4.22. *With the same conditions as in Theorem 4.21 but with* $P^\frown(\overline{B}_1 R_1) \neq 0$ *in place of* $P^\frown(B_1 R_1) \neq 0$ *then*

$$P^\frown[(\textstyle\bigwedge i)(B_i \dashv R_i) \dashv \overline{B}_1 R_1] = 0.$$

PROOF. As in Theorem 4.21, but with $\overline{B}_1 R_1$ instead of $B_1 R_1$.

$$P^*[\textstyle\bigwedge_{i=1}^{n-1}(B_{i+1} \dashv R_{i+1})(B_1 \dashv R_1) \dashv \overline{B}_1 R_1]$$
$$= P^*[(C \dashv A)(B_1 \dashv R_1) \dashv \overline{B}_1 R_1]$$
$$= P^*[(C \dashv A)(B_1 \dashv R_1)(\overline{B}_1 R_1) \dashv \overline{B}_1 R_1]$$
$$\text{Thm 1.52}$$

and since $(B_1 \dashv R_1)(\overline{B}_1 R_1) \equiv_u \mathbf{0}$,

$$P^*[(C \dashv A)\mathbf{0} \dashv \overline{B}_1 R_1] = 0.$$

□

Thus the supposition that there is a raven that isn't black suffices to make the $(P^\frown)$ conditional-probability, that everything that is a raven is black, equal to 0.

From a material point of view these results are not very substantial. All we have shown is that a confirming instance doesn't reduce the probability of a generalization (with suppositional) when adjoined as "evidence". But since no probability model was specified and only simple hypotheses such as $P(B_1 R_1) \neq 0$ and $P(\overline{B}_1 R_1) \neq 0$ was used, it isn't surprising. At least with the suppositional in place of the truth-functional conditional the paradox is dispelled.

In Theorem 4.21 we have seen how the generalization $\bigwedge_{i=1}^{n}(B_i \dashv R_i)$ fares with $B_1 R_1$ taken as a supposition, namely that it either confirms (adds probability to) it or else is neutral, in that there is no decrease in probability. It is of some interest to see what happens if instead of $B_1 R_1$ one takes $\overline{B}_1 \overline{R}_1$ as a supposition. For in the Hempel paradox $\overline{B}_1 \overline{R}_1$ is assumed to be as much an instance to be considered as $B_1 R_1$ since $(R_i \to B_i)$ and its contrapositive $(\overline{B}_i \to \overline{R}_i)$ are logically equivalent. This is the result:

THEOREM 4.23. *With the same conditions as in Theorem 4.21 but with* $P\tilde{}(\overline{B}_1\overline{R}_1) \neq 0$ *in place of* $P\tilde{}(B_1 R_1) \neq 0$, *then either*

$$P\tilde{}[(\wedge i)(B_i \dashv R_i) \dashv \overline{B}_1\overline{R}_1] = 0, \quad or \quad = c,$$

the latter value being the case if conditions specified at the end of the proof hold.

PROOF. Proceeding as in Case 2 of the proof of Theorem 4.21, but with $\overline{B}_1\overline{R}_1$ instead of $B_1 R_1$, one has

$$P\tilde{}(\wedge_{i=0}^{n}(B_i \dashv R_i) \dashv \overline{B}_1\overline{R}_1)$$
$$= P^*((C \dashv A)(B_1 \dashv R_1) \dashv \overline{B}_1\overline{R}_1)$$
$$= P^*((C \dashv A)(B_1 \dashv R_1)(\overline{B}_1\overline{R}_1) \dashv \overline{B}_1\overline{R}_1);$$

Thm 1.52

replacing $\overline{B}_1\overline{R}_1$ by $\overline{B}_1\overline{R}_1 \dashv \mathbf{1}$ and using §1.4(c), yields

$$= P^*((C \dashv A)(\mathbf{0} \dashv B_1 \vee R_1) \dashv \overline{B}_1\overline{R}_1),$$

and once again using §1.4(c),

$$= P^*((\mathbf{0} \dashv A\overline{C} \vee B_1 \vee R_1) \dashv \overline{B}_1\overline{R}_1)$$
$$= P^*(\mathbf{0} \dashv A\overline{C}\,\overline{B}_1\overline{R}_1) \qquad \text{§1.5(11)}$$

Thus, far from being a confirming instance, $\overline{B}_1\overline{R}_1$ either totally disconfirms $\wedge_{i=0}^{n}(B_i \dashv R_i)$, or else if $P(A\overline{C}\,\overline{B}_1\overline{R}_1) = 0$, assigns to it an indefinite value in $[0, 1]$. □

While one can't legislate what is to be the meaning of "a generalization" it seems that $(\wedge i)(R_i \to B_i)$ is less informative than $(\wedge i)(B_i \dashv R_i)$. By constructing semantic tables for $B_i \dashv R_i$ and $R_i \rightarrow B_i$ one has, with '$\Rightarrow_u$' standing for semantic entailment (see §1.4),

$$B_i \dashv R_i \Rightarrow_u R_i \rightarrow B_i$$

—if one knows that $B_i \dashv R_i$ holds then one also knows that $R_i \rightarrow B_i$ does, but not conversely. If the interest is in the color of a bird species it makes more sense to examine birds of that species, rather than objects in general, and '$\dashv$' with R_i as the antecedent of the suppositional does exactly that—it excludes everything but ravens.

We end our discussion with the following prescient conclusion of a lengthy article on the topic (*Black* 1966, p. 195):

(e) Conclusions. I believe that a *prima facie* case has now been made for thinking that the discomfort produced by the paradoxical cases of confirmation is partly due to the logical gap between material implication [truth-functional conditional] and "ordinary" implication. However it is hard to be sure of this in the absence of any thorough and comprehensive examination of the discrepancies between the two concepts.

BIBLIOGRAPHY

Barone, J. and Albert Novikoff

    1978   A history of the axiomatic formulation of probability from Borel to Kolmogorov. Part I, *Archive for history of exact sciences*, vol. 18, 123–190.

Black, Max

    1966   Notes on the "Paradoxes of confirmation", *Aspects of inductive logic*, J. Hintikka and P. Suppes, eds. (Amsterdam: North-Holland), 175–197.

Boole, George

    1857   On the application of the theory of probabilities to the question of the combination of testimonies or judgements, *Transactions of the Cambridge Philosophical Society*, vol. 11, 396–411. Reprinted as Essay XVI in *Boole* 1952.

    1952   *Studies in logic and probability*, R. Rhees, ed. (London: Watts).

Borel, Émile

    1909   Les probabilités dénombrables et leurs application arithmétiques, *Rendiconti del circolo matematico di Palermo*, vol. 27, 247–271.

Charnes, A. and W. W. Cooper

    1962   Programming with linear fractional functions, *Naval reserach logistic quarterly*, vol. 9, 181–186.

Couturat, Louis

    1903   *Opuscules et fragments inédits de Leibniz, Extraits des manuscrits de la Bibliothèque de Hanovre* (Paris: Presses Universitaires de France). Reprinted 1961 (Hildesheim: Olds).

Crocco, G., L. Fariñas del Cerro and A. Herzig

  1995   *Conditionals from philosophy to computer science.* Oxford Science Publications. (Oxford: Clarendon Press).

Dale, Andrew I.

  2003   *Most honourable remembrance.* The life and work of Thomas Bayes. (New York Berlin Heidelberg: Springer Verlag)

Dubois, Didier and Henri Prade

  1995   Conditional objects, possibility theory and default rules. Essay 10 in *Crocco, Fariñas del Cerro and Herzig* 1995.

Fréchet, Maurice

  1935   Généralisations des théorème des probabilités totales. *Fundamenta mathematicae*, vol. 25, 379–387.

Frege, Gottlob

  1879   *Begriffsschrift, eine der arithmetischen nachgebildete Formelsprache des reinen Denkens* (Halle).

Gaifman, Haim

  1964   Concerning measures in first order calculi, *Israel journal of mathematics*, vol. 2, 1–17.

Hailperin, Theodore

  1986   *Boole's logic and probability.* 2nd edition. Revised and enlarged. (Amsterdam New York Oxford Tokyo: North-Holland)

  1988   The development of probability logic from Leibniz to MacColl, *History and philosophy of logic*, vol. 9, 131–191.

  1996   *Sentential probability logic* (Bethlehem: Lehigh University Press and London: Associated University Presses)

  1997   Ontologically Neutral Logic, *History and Philosophy of Logic*, Vol. 18, 185–200.

  2000   Probability semantics for quantifier logic, *Journal of philosophical logic*, vol. 29, 207–239.

2006 Probability logic and combining evidence, *History and philosophy of logic*, vol. 27, 244–269.

2007 Quantifier probability logic and the confirmation paradox, *History and philosophy of logic,* vol. 28, 83–100.

2008 Probability logic and Borel's denumerable probability, *History and philosophy of logic*, vol. 29, 147–165.

Hempel, Carl G.
1945 Studies in the logic of confirmation, *Mind* vol. 54, 1–26 and 97–121. Reprinted in *Hempel* 1965.

1965 *Aspects of scientific explanation* (New York: The Free Press).

Herbrand, Jacques
1971 *Logical writings*, ed. W. D. Goldfarb (Cambridge, MA: Harvard University Press).

Kleene, Stephen C.
1952 Introduction to metamathematics (New York: D. Van Nostrand)

Kneale, William and Martha Kneale
1962 *The development of logic* (Oxford: Clarendon Press).

Kolmogorov, A.
1933 *Grundbegriffe der Wahrscheinlichkeitsrechnung* (Berlin: Springer).

Łos, J.
1955 On the axiomatic treatment of probability, *Colloquium math.*, vol. 3, 125–137.

Nicod, Jean
1930 *Foundations of geometry and induction*, trans. by P. P. Weiner (New York: Harcourt).

Renyi, Alfred
1970   *Probability theory* (Amsterdam: North Holland and New York: American Elsevier).

Saunders, Sam C., N. Chris Meyer and Dane W. Wu
1999   Compounding evidence from multiple DNA-tests, *Mathematics magazine*, vol. 72, 39–43.

Scott, Dana and Peter Krauss
1966   Assigning probabilities to logical formulas, *Aspects of inductive logic*, J. Hintikka and P. Suppes, eds. (Amsterdam: North Holland), 219–265.

Shiryayev, A. N.
1984   *Probability* (New York: Springer).
1995   Second edition

Smorynski, C.
1977   The incompleteness theorems, *A handbook of mathematical logic*, Jon Barwise, ed. (Amsterdam: North Holland), 821–866.

Suppes, Patrick
1966   Probabilistic inference and the concept of total evidence, *Aspects of inductive logic*, J. Hintikka and P. Suppes, eds. (Amsterdam: North Holland), 49–65.

Swinburne, Richard
1973   *An introduction to confirmation theory* (London: Methuen).

Von Plato, J.
1994   *Creating modern probability* (Cambridge UK: Cambridge University Press).

INDEX

122

www.ingramcontent.com/pod-product-compliance
Lightning Source LLC
Chambersburg PA
CBHW021426180326
41458CB00001B/151